The Regent Diamond

The Logan Sapphire

The Hooker Emerald

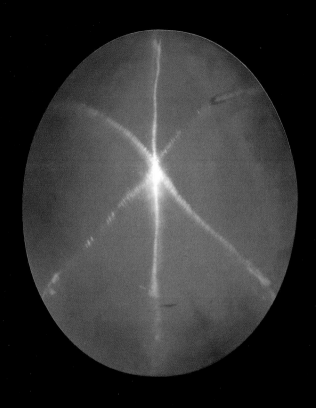

The Rosser Reeves Star Ruby

CONTENTS

DIAMONDS AND PRECIOUS STONES

Patrick Voillot

DISCOVERIES®
HARRY N. ABRAMS, INC., PUBLISHERS

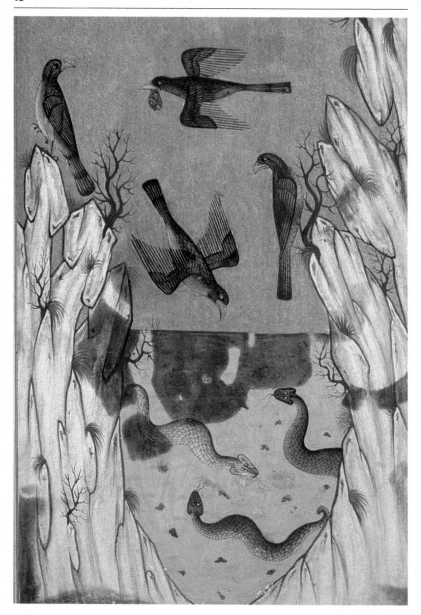

"Was it a flash of divine insight, or the slower process of observation and deduction, that led human beings to perceive [an] esoteric quality in stones? They saw beauty in the sunrise; but the sun became blinding by midday. There was color in leaves and flowers, until they withered. Water sparkled, but it could not be worn for long. Of all the natural wonders of the earth, only the stones endured. They must indeed be magical; and those who possess magical things can sometimes put to work the magic in them."

Cornelia Parkinson,
Gem Magic: The Wonder of Your Birthstone, 1988

CHAPTER 1

THE MYSTERIOUS ORIGINS OF PRECIOUS STONES

Left: the legendary Valley of Diamonds, guarded by snakes with deadly eyes. Right: an emerald fresh from the mine, still embedded in its calcite matrix.

Nature's marvels

Since the beginning, humankind has been fascinated by gems—minerals pleasant to touch and filled with mystery, whose shapes captivate and whose colors and luster entrance. They enthralled our ancestors, partly because of their exceeding rarity, and they loomed large in the symbolism of the ancient world. Nor have they lost that magical charisma.

In Western culture, the symbolic meanings of gemstones were traditionally related to their colors, which in turn were associated with moral qualities through imitative or sympathetic magic. According to common belief, a stone's color also gave it the ability to heal body organs of the same shade: thus, a ruby was thought to cure disorders of the blood, an emerald those of the eyes, and so on. Even today some patent medicines contain powdered gems, and with the renewed interest in alternative medicine that has arisen at the end of the 20th century lithotherapy, the use of gems and crystals for healing purposes, has enjoyed a new popularity.

A gemstone's color conveys its symbolic power. Inspired by nature, our ancestors chose the white brilliance of the diamond to represent light, the green of the emerald for the rebirth of the seasons and of life, the red of the ruby for fire, and the blue of the sapphire for the heavens. From the beginning, the world's great religions have mentioned gemstones in their sacred texts as important symbols of spiritual values. Precious and extremely

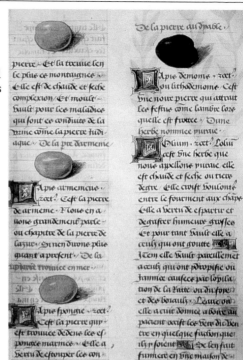

A 15th-century French treatise lists the curative powers of lapis, pumice, and a mineral called devil's stone, described as a black stone that can take a static charge, probably jet. Similar beliefs are found also in Ayurvedic medicine in India, and even in the artifacts of ancient Mesopotamia, recorded on clay tablets listing stones that protect against various maladies when worn around the neck.

rare, these stones were seen as having an exceptional, almost divine, nature. This quasidivine quality was reinforced by the fact that until the 15th century most gems were mysterious to Europeans. Their origins were in distant, unknown lands; they arrived from these exotic, half-mythical places almost miraculously, carried by merchants or messengers from the ends of the world, and came to their destinations bereft of information about the place or manner of their extraction.

According to the Old Testament (Exodus 28), the ancient Israelites chose twelve precious stones to ornament

B elow: Aaron, high priest of the Jews, wears his jeweled breastplate in a 15th-century illustration. Left: a panel from Aaron's jeweled pectoral, as imagined in a detail from a 15th-century illuminated manuscript.

the "Breastplate of Judgment" of the high priest Aaron, each engraved with the name of one of the sons of Jacob, whose descendants constituted the twelve tribes of Israel. The emerald stood for Ruben, the sapphire for Dan, and the ruby for Naphtali. It is noteworthy that the diamond was not included (the word *yahalom*, commonly translated as "diamond," probably means "jade"). In the New Testament (Rev. 21:18–20) the twelve foundations of the walls of Jerusalem are envisioned as adorned with precious stones, of which the emerald and the sapphire symbolize the moral virtues of the holy city. In Islam, the ruby in the Kaaba, the great shrine at Mecca, is revered as the stone of the Last Judgment.

The hierarchy of stones

It was not until the Middle Ages that a real art of precious stones emerged in Europe with the craft of religious and royal jewelry making, which achieved marvels of refinement in the 14th and 15th centuries and reached even greater heights at the end of the 16th. The fantastically rich banking families such as the Fuggers and Welsers of Augsburg and the Medici and Strozzi of Florence amassed fabulous private collections of precious stones, often in spectacular settings. In the history of humankind diamonds have not always been thought of as the most valuable stones, as they usually are today. In antiquity turquoise, lapis lazuli, amethyst, jasper, and cornelian were preferred; China and pre-Columbian Mexico loved jade; Rome, the sapphire and the emerald; diamonds and rubies were rare. Indeed, these last four gems all had to wait until the 20th century for official description and recognition as precious stones.

The unconquerable diamond

The diamond, now held to be the preeminent precious stone, was not always esteemed so highly. It was not until the first century AD that the Greek word *adamas*, unconquerable, appeared in Pliny the Elder's *Natural History* to describe it. Pliny, however, did consider this stone the most precious of worldly goods. The diamond, which owes its name to its hardness, was valued in ancient times for its physical properties of resistance to shock and fire. To it as well was attributed power for both good and evil; it was thought to dissolve in goat's blood and to destroy the lodestone's capacity to attract iron.

A bove: this Siren pendant, made in Germany around 1580–90 and ornamented with rubies and pearls, was once in the collection of the Medici family, rulers of Renaissance Florence. In that period, according to the goldsmith and sculptor Benvenuto Cellini (1500–1571), a ruby weighing 1 carat was worth twice an emerald of the same weight, and eight times more than a diamond.

S tones considered merely semiprecious today have, in other times, been highly valued. Left: a pre-Columbian earring in gold and lapis lazuli.

The Greeks did not use the diamond for adornment and the Romans did so only occasionally, mounting it uncut in its natural form, an octahedron, or eight-faced shape. At all events, it was so rare that treatises were written describing ways to fabricate fake diamonds that could compete with real stones; these manuscripts so angered the 3d-century AD emperor Diocletian that he

A group of 16th-century paintings in the city hall of Florence describes the Medici taste for gems. Above: a jeweler's workshop; following pages: scenes of imaginary diamond and gold mines.

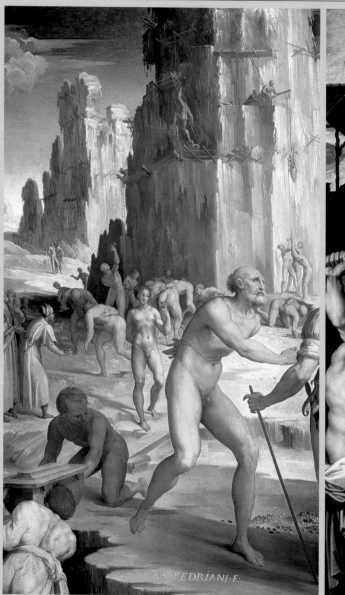

ordered them all burned. What was rare was costly, and should remain so.

Though the diamond was rarely an ornamental stone, engravers in the Greco-Roman world and in China made use of its extreme hardness: mounted at the tip of an iron tool called a burin, it was used as an instrument to engrave cameos. This practice, which dates to the 2d century BC, is mentioned in a Chinese manuscript from that period, the *Lie Tseu,* or *Book of Master Lie.*

Because of its remarkable hardness and clarity, the diamond attracted innumerable superstitious beliefs, some of which persisted until the 17th century—or even longer in some places. The stone was credited with virtues of purification and invincibility, or even the ability to create other diamonds spontaneously. It was supposed to calm anxiety, give protection from ghosts, heal cataracts. In Arab folk medicine the diamond, the stone of perfection, was thought to cure all the ills of the body and mind.

The legend of the Valley of Diamonds

The symbolic importance of gemstones has given rise to countless stories and tales about them. That of the Valley

B elow: the fable of the Valley of Diamonds is illustrated in a 1375 atlas. Right: a modern illustration from *Oriental Tales.*

of Diamonds, which dates to the 4th century BC, has been repeated almost unchanged for centuries, and appears in a number of folktales and fantastic images. According to this legend, somewhere in the Scythian desert to the north of the Black Sea there once was a deep, inaccessible gorge, guarded by ferocious snakes and eagles, in which lay a fortune in diamonds. A mighty ruler sent his terrified servants on an impossible quest to collect these stones. They eventually thought up a plan to avoid coming back empty-handed to face the king's wrath: they killed sheep and cut them into quarters, which they threw to the bottom of the gorge. The eagles spotted the bait and flew down into the chasm to gather and carry away the pieces of meat, to which the gems stuck. The king's men had only to plunder the raptors' nests to retrieve the precious spoils.

The legend of the Valley of Diamonds contributed to the reputation of the white gem as difficult to find and perilous to acquire. Below: an illumination from a 15th-century French edition of the *Book of Marvels,* which recounts the adventures of Marco Polo. A mid-5th-century treatise from India, the *Buddhabhatta,* maintained that whoever wore a diamond would have a life free from all dangers, even when faced with snakes, fire, poison, illness, thieves, water, and evil spirits. Thanks to these powers, it was said, Indian merchants captured the market for diamonds in Rome, where the stone was valued as a talisman.

First mentioned in the Middle Ages, the Emerald Table of Hermes was supposed to be engraved with the precepts of alchemy, allowing alchemists to create the philosopher's stone. This arcane jewel symbolized the ultimate secret that all seekers after arcane knowledge desired. Left: a monk examines the Emerald Table, in an illumination of the 14th or 15th century from the *Aurora Consurgens.*

Emerald, the alchemist's stone

The emerald, a type of beryl, takes its name from its color—*smáragdos* means "green stone" in Greek—and symbolizes the life force of the cosmos, strength, and inner bliss. In Central America it was a fertility token. The Aztec called it *quetzlitzli,* associating it with the quetzal, a mythical green-feathered bird, symbol of spring's renewal. It was said in India that the mere sight of an emerald would cause a viper or cobra such terror that its eyes would leap out of its head.

Esoteric lore held that a magic tablet made from an enormous emerald had been discovered in ancient Egypt, buried with the mummy of Thoth, the Egyptian god of knowledge. It was said to be engraved with the precepts of the occult sciences; the so-called Emerald Table long remained a favorite fable in the history of mythology. Twenty centuries before Christ, in the Middle Kingdom of ancient Egypt, emeralds sometimes ornamented the funerary jewelry of the nobility. They were exceedingly rare, so green pottery or colored glass was often used as a substitute.

Emeralds were held to have all sorts of medicinal properties. Ancient texts often speak of their calmative powers. It is said that the 1st-century AD Roman emperor Nero watched gladiator combats through emerald crystals to quiet his excitement at the sight of that cruel sport, and perhaps because they were said to sharpen sight. The 16th-century French writer François Rabelais also reminds us of their supposed aphrodisiac powers when he gives his licentious character Gargantua a pair of pants fastened with two hooks trimmed with emeralds "the size of an orange." For sailors, fishermen, and other seafarers the emerald is among the most powerful talismans, granting protection from tempests by association of the stone's serene green color with that

In Peruvian tombs from the Chimu culture (12th–14th century), masks like this one, which has strings of emeralds in place of the eyes, were laid on the folded clothes that accompanied the deceased. It was long thought that the emeralds adorning these Peruvian funerary objects were of local origin. In fact, they came from Colombia, the foremost producer of emeralds in South America.

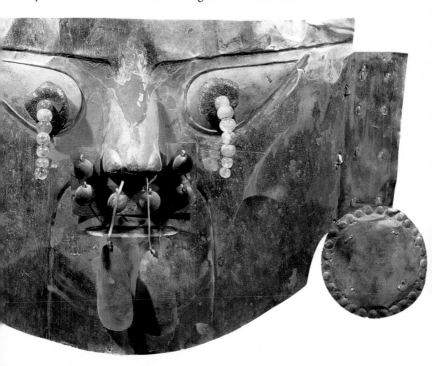

of the calm sea. According to an alchemists' legend, the philosopher's emerald is of all stones the one that breathes life into nature.

As is often the case with the symbolism of gemstones, positive qualities can be transposed into negative ones, symbolizing peril and danger. In Christian symbolism the emerald is associated with the most dangerous creatures of Hell, and in the Middle Ages especially it sometimes represented evil: in the Munich Residenz-museum is a gem-studded reliquary from 1590 in which a ruby-bedecked Saint George strikes down a Dragon glittering with emeralds. The Holy Grail, too, was said to have been cut from an emerald that fell from Satan's helmet. Medieval folklore thus granted the emerald both dangerous and beneficial powers, in which a hint of witchcraft is necessarily mixed. Originating in Hell, the green stone was able to destroy infernal creatures, whose secrets it knew.

Although the Prophet Muhammad forbade the faithful to covet worldly goods or wear expensive jewelry, Muslim sovereigns particularly loved precious stones. The caliph Hārūn ar-Rashīd (763–809 AD), ruler of the great Arabic empire whose capital was Baghdad, used to send his jeweler on missions to Sri Lanka to acquire stones for his private collection. Above: an illumination from Mîr Haydar's 15th-century Turkish book *Mi'radj Nâmeh,* depicting Muhammad at the foot of the Tree of the Emerald Branches.

Ruby, fire of passion

The ruby, a corundum stone, owes its name to its red color (*rubeus* in Latin; the name of the family of ruby and sapphire comes from the Hindu word *kurand* or *korund*). It is the archetypal color of fire and symbolizes daring, charity, and divine love. In India, Hindu tradition has always considered it the most valuable of stones; in Sanskrit it is called *Ratna Raj,* "queen of precious stones."

Philippe de Valois, in his treatise *Lapidary,* calls it the prince of gems, and the Italian Renaissance poet Francesco Petrarch tells us that in the Middle Ages John the Good, king of France, always wore a ring made of several rubies as a good-luck charm. Sovereigns wore them too because of their prophetic power: they were said to announce evil omens by suddenly turning from pure red to black.

We find a similar belief in Islamic tradition. The ruby

The Hindu god Vishnu, incarnated as a fish in this Indian miniature of the 18th-century Pahari school, bears on his breast a magical jewel traditionally described as a shining ruby. In classical Hindu texts many kinds of red stones were called rubies, but with the appearance of scientific techniques for identifying gems some of these have proved to be semi-precious spinels or garnets. True rubies from the ancient world are very rare.

of Abraham, brought to the Kaaba by the angel Gabriel, was turned black by the sins of mankind. This legendary stone is said to have eyes, ears, and tongue, so that it may testify on behalf of the righteous on the day of the Last Judgment. This ruby is the holy stone of Islam.

A jewel representing both religious and secular power, the ruby has often appeared prominently in the royal crowns of Christian monarchs in Europe, in remembrance of the suffering and blood of Christ. The stone thus symbolizes the sacrifice of the sovereign, who puts himself at the service of his or her country and people. The crown of

B elow: the Crown of the King of Bohemia is set with a 250-carat ruby and other precious gems.

Saint Wenceslas, king of Bohemia in the 10th century—so called because it comprises a reliquary containing the saint's skull—is mounted with a jewel that is probably the largest gem-quality ruby known. It measures 1.55 by 1.44 by 0.55 inches (39.5 x 36.5 x 14 mm) and weighs 250 carats. (The carat, which equals 200 milli-grams, is the unit of measurement used to weigh precious stones.) This crown was commissioned by Charles IV of Luxemburg in 1346, and is embellished with numerous precious stones and four fleurs de lys, a reference to France, where Charles grew up.

Opposite: a ruby ornaments the turban of the 15th-century Romanian ruler Vlad Tepeş, known as Vlad the Impaler, a relative of the original Dracula. Left: this is the pendant known as the Talisman of Charlemagne, 9th-century Holy Roman Emperor and king of the Franks. It was a gift from the Arab caliph Hārūn ar-Rashīd. The central stone is a big cabochon-cut sapphire set in a gold mount with gems and pearls. Charlemagne bore the jewel to his tomb in 814 AD, and legend accorded it miraculous powers. This belief was reinforced when Otto III (980–1002 AD) had the tomb of the emperor opened and found the body in a remarkable state of preservation. The talisman was added to the treasury of Aix-la-Chapelle, Charlemagne's city, and in the 19th century was presented to Empress Josephine, consort of Napoleon. It then passed to Princess Eugénie, wife of Napoleon III, who gave it to the city of Rheims to support the reconstruc-tion of the cathedral there, which had been damaged in World War I.

Sapphire, the celestial stone

The sapphire, a blue corundum, belongs to the same mineral family as the ruby. It is emblematic of immor-tality and chastity. It represents the spiritual values of humanity and has always stood for the mystery of the celestial. The Persians believed that the earth rested on a giant sapphire, which radiated its magnificent color throughout the universe. The Egyptians and the Romans revered the sapphire as the stone of justice and truth.

The Catholic church chose it as the supreme symbol of the light of God, and bishops and cardinals were advised to wear a sapphire on the right hand as a token of the powers of Heaven, grantor of their authority to administer blessings and judgments. This custom was recorded in a papal bull issued by Pope Innocent III in the 12th century, and persisted until the 14th century among English bishops, who received a sapphire ring on the day of their elevation.

The creation of precious stones

Like true works of art, gemstones are the fruit of an age-long gestation in the earth's bowels. Their formation requires unimaginable pressures and temperatures; one might almost say that they are the creation of time itself, in whose patient embrace matter re-forms and crystallizes into glowing baubles.

The diamond, one of the oldest minerals in the universe, is made of pure carbon. It forms at a depth of from 90 to 120 miles beneath the earth's surface, at temperatures above 2200° F (1200° C). Usually brought to the surface from its primary deposit by volcanic action, in lava, it is found embedded in kimberlite, a volcanic rock named after Kimberley, the South African town where it was first identified. Erosion then frees it and carries it to its secondary deposit in river- and seabeds.

Modern dating techniques have allowed us to approximate the age of one of the biggest diamonds, the Cullinan, at between 1.6 and 1.7 billion (thousand million) years; more recently formed diamonds may be 70 million years old. Traces of diamond with an unusual

Inclusions are common flaws in precious stones. Parasitic inclusions are often given descriptive names: silk, feathers, horsetails, treacle. In diamonds they are considered undesirable because they block the passage of light, but they do not always reduce the quality and worth of an emerald. Clustered mossy inclusions are typical of fine Colombian emeralds. Below: a crystal inclusion in one such.

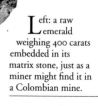

Left: a raw emerald weighing 400 carats embedded in its matrix stone, just as a miner might find it in a Colombian mine.

crystalline structure have been discovered in the hearts of extraterrestrial meteorites that have fallen to earth.

Sapphires and rubies are made of aluminum oxide and are formed, like diamonds, in the earth's deep-subsurface lithosphere. Their crystallization, however, takes place at a shallower depth, as does that of the emerald, created through contact between pegmatites (igneous granites) and basic rocks from deep within the earth, the source of the chrome which gives it its green color. The pegmatites of Brazil, source of many famous emeralds, are between 450 and 490 million years old. The ruby owes its red color to the presence of chrome oxide in its composition.

Far from lowering a gem's price, inclusions may guarantee its authenticity and provenance, for natural inclusions are catalogued by type for the four precious stones and may serve to identify a famous jewel. The interior of every ruby, for example, tells the tale of its crystallization and the geological conditions of its formation. Inclusions of mica, zircon, other microscopic minerals, or liquids inform us about the geologic and thus geographic origin of a stone. Liquid inclusions look like feathers in the shape of a fan in Burmese sapphires, while in those from Sri Lanka they are shaped like butterflies' wings. Left: a cut ruby from Thailand with an inclusion of eroded crystal.

Inclusions

Precious stones often contain foreign bodies, locked within their natural crystalline structure, more or less visible to the naked eye. These may occur in solid, liquid, or gaseous form and can be studied with a strong lens. Burmese rubies contain interlaced needles of rutile (a mineral), which look like woven silk when examined under a microscope. Finer rutile needles may also be found in Ceylonese rubies and in sapphires.

"The first piece which Akil Khān placed in my hands was the great diamond, which is a round rose, very high at one side.... After I had fully examined this splendid stone,...he showed me another stone, pear-shaped.... All these stones are of first-class water, clean and of good form, and of the most beautiful kind ever found."

Jean-Baptiste Tavernier,
Travels in India, 1681

CHAPTER 2
PATHS PAVED WITH GEMS

Left: Shāh Jahān (r. 1628–58), Mughal emperor of India, sits on the Peacock Throne, made entirely of gold and gems, a symbol of the immense wealth of his court. Right: Hernán Cortés, the Spanish conquistador, is given an emerald necklace by the Aztec emperor Montezuma.

The Koh-i-Noor: a fairy-tale diamond

The Koh-i-Noor, or "Mountain of Light," a name that
dates to the 18th century, is not the biggest diamond
among the British crown jewels—although you might
think so to see the bulletproof glass that protects it in the
Tower of London—but thanks to its legendary history it
is probably the most famous diamond of all time.

In India it is said that the Koh-i-Noor was discovered
on the forehead of an abandoned infant on the banks of
the Yamuna River, and that a mahout's daughter took in
the baby and brought him—and the stone—to court.
The child was none other than Karna, son of the Vedic
sun god. The stone, originally 600 carats of crystallized
light, was fastened to a statue of Shiva over the third eye,
the eye of enlightenment.

In 1304 the stone is mentioned for the first time in
a written chronicle as the property of the rajah of
Malwa. Then there is no mention of it for more than
two centuries, until it reappears in the treasury of
Muhammad Zahīr-ud-Dīn (called Babur, 1483–1530),
founder of the Mughal dynasty in 1526.

The Mughals possessed the stone for 200 years, until
Nāder Shāh (1688–1747), sovereign of Persia, sacked Delhi
in 1739. The fabulous diamond was missing from among
the spoils of war, the conquered ruler, Muhammad Shāh,
having hidden it in the folds of his turban. In observance
of an ancient custom, the conqueror held a feast for the

Above: the Persian
conqueror Nāder
Shāh asks Muhammad
Shāh to give up the
turban in which the
Koh-i-Noor is hidden.
The diamond was
passed down to various
sovereigns; the last, Ranjit
Singh (1780–1839), the
Lion of Punjab, founder
of the Sikh kingdom, was
forced to give it to the
British.

vanquished leader at which the two exchanged turbans as a token of peace; Nāder Shāh thus captured the greatest prize of his triumph. After his assassination in 1747, his son inherited it; it is said that he chose death by torture rather than give up the fabulous diamond. It then passed into the hands of other owners, Afghans and Sikhs (who ruled the kingdom of Punjab), before being spirited away by the English when they entered Lahore in 1849.

On the 250th anniversary of the creation of the East India Company, the stone, by now cut to 186 carats, was presented as a gift to Queen Victoria. It traveled to London aboard the ship *Medea,* hedged about with safeguards, and was put on display in the Crystal Palace at the Universal Exposition of 1851 in London. It did not cause much stir, as its Indian cut gave it only a dull shine. The queen summoned the famous diamond cutter Voorsanger of the Amsterdam jewelry house of Coster to recut the Mountain of Light. His efforts further reduced the diamond's weight to 108.93 carats, but much enhanced its brilliance and gave it an international fame that has never faded.

Top: the Koh-i-Noor diamond is now mounted in the crown of Queen Elizabeth the Queen Mother. Above: the recutting of the stone in the 19th century was an inspiration for English cartoonists. Prince Albert placed the stone on a steam-driven machine and the Duke of Wellington started it up. On the first day of the Universal Exposition Queen Victoria is supposed to have said, "This is the best day of my life."

No one will ever know the true story of the discovery of the Koh-i-Noor, nor the date or place where it was first cut. It is nevertheless very probable that it came from the mines of Bijapur, in central India, the only country in the world to produce diamonds prior to the 18th century. This Indian production of diamonds was a great mystery to Westerners until the Frenchman Jean-Baptiste Tavernier traveled to India in the 17th century and brought back the first reliable information about the diamond mines.

LES SIX
VOYAGES
de Jean Baptiste Tavernier
Ecuyer, Baron d'Aubone
qu'il a fait
en TURQUIE en PERSE,
et aux INDES,
pendant quarante ans
NOUVELLE EDITION
revüe et corrigée.
1712.

Left: the frontispiece of Tavernier's book. Tavernier (below) visited three mines, Raolconda (probably Golconda), Gani Coulour (Kollur), and Soumelpur.

Jean-Baptiste Tavernier's Indian travels

Tavernier, the son of a map merchant, was born in Paris in 1605 and made his fortune by trading in precious stones. Making six trips to India and Persia between 1631 and 1668, in an age when these lands were still a great mystery to Europeans, he acquired a great number of exceptional gems—the most splendid of which were sold to Louis XIV—as well as the extraordinary title of official trader for the French crown.

Tavernier saw the treasury of the Mughal emperors and visited the legendary diamond mines of Golconda in the south of India, documenting these with remark-able written descriptions. The first reliable information we have about diamond mining comes from his account. Nothing of the place remains today but the ruins of the fort that was the official residence of the sultans and maharajahs. But Golconda was the place where the most beautiful gems of Asia were once bought and sold. Tavernier tells us that the local princes governed the mines, keeping the biggest diamonds for themselves and forbidding the opening of too many sites, lest the greed of neighboring kingdoms be aroused. There were twenty-three mines in the kingdom of Golconda in 1678, many of which, as Tavernier notes, owed their discovery purely to chance: "It was by means of a poor man," he recounts in *Travels in India,* "who, digging a piece of ground where he purposed to sow millet, found a [raw diamond] weighing nearly 25 carats. This kind of stone being unknown to him, and appearing to him something special, he carried it to Golconda, and by good luck addressed himself to one who traded in diamonds." This was the beginning of the Kollur mine. "The noise of this new discovery quickly spread abroad, throughout all the country, and some persons of wealth in the town commenced to mine in this land.... There are found here at present, I say, a quantity of stones from 10 up to 40 carats, and sometimes indeed much larger; but among others the great diamond which weighed [90 carats] before cutting, which Mîr Jumla presented to [Shāh Jahān], as I have elsewhere related, was obtained from the mine."

Historical sources, though rare, establish the fact that in the 17th century mining was not limited to working the relatively accessible alluvial and fluvial deposits. Miners also attacked the rocky strata, digging tunnels and shafts. These mines were limited by the available technology of the time. Where possible, miners dug quarries before sinking shafts, which could reach a depth of 80 feet (25 meters). However, frequently they were content to scrape holes a few yards or meters in diameter and not more than about 20 feet (6 meters) deep. Tavernier tells us that men did the heavy work, while women and sometimes children sorted and examined the quarried stones. The diamonds, he wrote, were found in rocky fissures filled with silty earth: "Each commences to

Precious stones and jewels are listed in this 18th-century inventory of the treasures of Emperor Akbar of India (r. 1556–1605). Akbar's rule was famed for the brilliance of its arts and architecture and was a political and artistic apogee of the Mughal dynasty. Jade, wood, mother-of-pearl, gold, and precious stones were crafted in the workshops of the royal palace.

work, the men to excavate the earth, and the women and children to carry it to the place which has been prepared…. Men, women, and children raise the water with pitchers from the hole which they have excavated, and throw it upon the earth which they have placed there, in order to soften it…. [They] let off the water, then they throw on more, so that all the slime may be removed, and nothing remain but sand…[which] then they leave…to be dried by the sun…[and] which they agitate as we do when winnowing grain. The fine part is blown away…. All the earth having been thus winnowed, they spread it with a rake and…search for the diamonds."

A connoisseur, a polyglot, and an indefatigable traveler, Tavernier crossed and recrossed Europe. He was a brilliant trader, with a reputation for integrity; he usually traded little Italian jewels of no great value, coveted for their beauty by the Indians, for superb stones. He owed his success to his skill at adapting to a foreign culture; he often appeared wearing opulent silk turbans, or wrapped in rich furs.

Among the stones he acquired on his sixth trip, in 1669, was a blue diamond of superb beauty, 112 carats rough, which he named the Tavernier Blue and sold to Louis XIV, king of France, for 3 million *livres*. It was called the Great Blue Diamond at the French court. Louis is said to have worn it only once; Louis XV loaned it to his mistress, Madame du Barry; Louis XVI gave it to Marie Antoinette to wear, until he had it mounted in a magnificent ornament for the Order of the

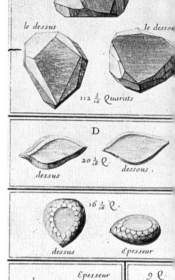

REPRESENTATION de Vingt des pl[...] au ROY, a Son dernier retour des Ind[...] consideration, et des Services que ledit[...]

l'Épesseur

le dessus

le desso[...]

A

112 3/10 Quarats

D

20 1/18 Q.

dessus

dessous.

16 1/18 Q.

dessus

Épesseur

Épesseur

dessus

9 Q.

10 1/2 Q.

Le DIAMENT cotté A, est net et d'vn beau Violet.
Ceux cottéz B, et C, Sont de couleur de rose-pâle. Celuy cotté D, est d'vne Eau extraordinairement belle.

Above left: Tavernier bought the Great Blue Diamond from the Kollur mine and sold it to Louis XIV. The diamond was subsequently recut, which reduced its weight to 67 carats.

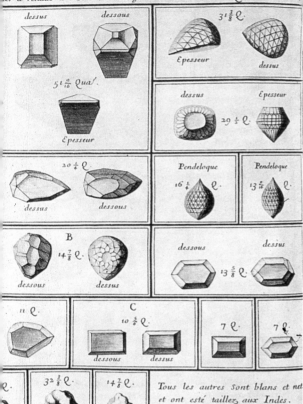

DIAMENS choisis entre tous ceux que le S.ᵗ I.B. Tauernier a Vendus esté le 6.ᵉ Decembre 1668. ou il a fait Six Voyages par terre, Et en cett ...ier a rendus a l'Estat, Sa Majesté la honnoré de la Qualité de Noble.

dessus

dessous

5 1 2/16 Quaʳ.

Epesseur

3 1 3/8 Q.

Epesseur

dessus

dessous

Epesseur

29 1/2 Q.

dessus

Epesseur

20 1/4 Q.

dessus

dessous

Pendeloque

16 1/4 Q.

Pendeloque

13 1/4 Q.

B

14 7/8 Q.

dessous

dessus

dessous

dessus

13 5/8 Q.

dessus

11 Q.

C

10 2/4 Q.

dessous

dessus

7 Q.

7 Q.

... Q.

32 3/8 Q.

14 7/8 Q.

Tous les autres sont blans et net et ont esté taillez aux Indes. Les trois d'Embas Cottez, 1, 2, 3. sont Bruts.

II. Partie. fol. ...

Golden Fleece. It was stolen in 1792 in a famous burglary of the treasury, and reappeared mysteriously in London in 1830, where it was bought by the banker Henry Philip Hope, who gave it its present name. The stone continued to travel: it turned up in the safe of a New York jeweler, then in that of a Russian prince, and finally in the treasury of an Ottoman sultan, before it was bought by an American named McLean. When some of its

L eft: in the 17th century the crown of France acquired many precious stones from Tavernier. This engraving from his book shows the diamonds he sold to Louis XIV (1638–1715), among them the Great Blue, later called the Tavernier Blue, and then the Hope Diamond (seen uncut at the bottom of the chart). The subject of fantastic adventures and the property of famous owners, the blue diamond was set in a pendant (below) by the jeweler Pierre Cartier in 1911 and sold to the American billionaire Edward B. McLean, owner of the *Washington Post*, for the sum of $15,000 (£9,000).

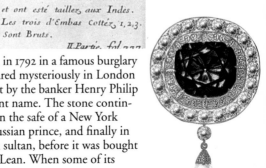

owners were said to have suffered an unhappy, even tragic fate, it gained a reputation for being cursed. In 1958 its last owner, the American jeweler Harry Winston, gave it to the Smithsonian Institution in Washington, D.C.

As for the illustrious Tavernier, he was bankrupted by his son. According to one tale, he once again left for India, where he was devoured by wild animals; another relates that he died a natural death in Muscovy at the age of more than eighty.

Cleopatra's mines

Most of the emeralds known to the world from the first millennium BC until the discovery of the Americas were drawn from secret mines established in pharaonic Egypt. References to these legendary deposits are few; one tells us that in the 14th century BC Seti I (r. 1318–1304 BC) went to the Western Desert to look for gold and emeralds.

In the early 19th century a Frenchman named Frédéric Caillaud explored the desert on the shores of the Red Sea, near the mountain called Gebel Zabara. There he discovered the site of what were probably the mines of the Egyptian queen Cleopatra (69–30 BC), whose favorite stone was the emerald. The area was riddled with galleries about 80 feet (25 meters) deep, piercing strata of black mica schist. Tools were found there dating to 1333 BC. These mines, which never produced any exceptional emeralds, were intensely active during the Greco-Roman period, and again in the Ottoman era, until about 1740, when they were abandoned for good. At the beginning of the 20th century several attempts were made to revive their magnificence, but all ended in failure.

Egyptian emeralds circulated throughout the Mediterranean, although only a few very rare specimens remain in museums today. They were especially prized by the Romans, who attributed curious powers to them. Tradition has it that the magnificent emerald which adorned the tiara of Pope Julius II (1443–1513), called "the

These pendant earrings originated in Egypt. They are made of gold, set with cameos and emeralds, and date from the 3d century AD, when Rome ruled North Africa.

Frédéric Caillaud was the first Westerner to visit the famous emerald mines of Cleopatra. He illustrated his itinerary in his book, *Voyages à l'oasis de Thèbes* (*Voyages to the Oasis at Thebes*), published in Paris in 1821. This map of the route he took through the desert between the Nile and the Red Sea shows the mines and the town of Sekket.

Until the Spanish exploration of the Americas in the 16th century, Egyptian emeralds were the most famous in the Western world. Nevertheless, few of the stones in ancient jewelry are of high quality; although they are a bright green color, they are often very frosty. Below: a ring for two fingers from the 1st century BC, adorned with an emerald from Cleopatra's mines.

Terrible" for his severity, came from Egypt. In reality the gem's source is Colombia, and it was presented to Pope Gregory XIII (1502–85).

The fabulous emeralds of the Americas

The world's shipping lanes are strewn with astonishing wrecks that could tell amazing tales of the trade in precious stones. In 1993 a team of professional divers directed by Victor Benilous organized a plan to salvage wrecks of archaeological value off the East Coast of the United States. Archaeologists had found information about an 18th-century wreck of particular interest in the log of another ship that had sailed through the same waters in 1756, and had mentioned sighting a sailing ship in flames.

Twelve nautical miles off the Florida coast, in deep water not far from Cape Canaveral, instruments recorded three Spanish anchors dating to the colonial period. The divers got to work and discovered one of the most stunning archaeological finds of the century. They brought up a booty of incalculable worth, including a skull cut from rock crystal, 25,000 carats of cut emeralds, another 25,000 carats of polished emeralds, a mass of emerald crystals weighing 24,644 carats, jewelry, and hundreds of pieces of exceptional jewelry of Aztec and Mayan origin, some of them pre-Columbian ceremonial objects of solid gold.

But the prize of prizes was a treasure that had been thought lost for good, the Queen Isabella Emerald, an unusually shaped, oblong, unset stone that had belonged to Hernán Cortés (1485–1547), the Spanish conqueror of Mexico. The emerald weighs 964 carats and is too big to hold in a closed fist. Cortés named it for the queen of Spain, who died in 1504, the year he set sail for the New World. He sent it as a wedding gift to his second wife, Doña Juana de Zuniga, who sailed to Mexico and the West Indies with her husband, and who lived there with her family. Cortés had amassed a quantity of precious objects for her, notably perfume bottles cut from large emeralds, and above all this exceptional gem. Two hundred

Left: included in the New World treasure discovered by Victor Benilous is this gold cross from the colonial period, with seven cabochon emeralds; it measures 3⅛ x 2 inches (7.9 x 4.9 cm). Below: the famous Queen Isabella Emerald, a legendary stone, was believed to be lost forever. Of exceptional clarity, it has been appraised at more than $20 million (£12 million).

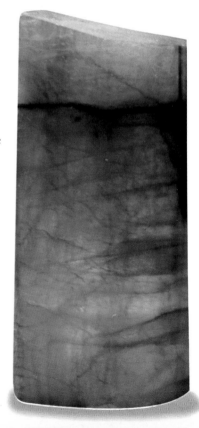

years later, the descendants of the Zuniga family chartered a little sailing ship to send these treasures back to Spain; it sank in the waters off Florida.

Cortés owned other unique emeralds. Several sources report that he always wore a necklet of five emeralds engraved with flowers, birds, and a little bell, a gift

to him from the Aztec emperor Montezuma. It is assumed that this jewel was lost in a shipwreck off the coast of Algiers, when Cortés battled Barbary pirates in 1541.

Chivor and Muzo, El Dorados of the Indians and conquistadors

Cortés was not the first European to pillage the emeralds of the Central and South American kings; before him the Spanish explorer Pedro Arias Dávila (1440?–1531) had obtained several beautiful stones, after landing on the Caribbean coast of Colombia, near what is now Santa Marta.

The first deposits in Colombia ever worked by Europeans were discovered by Captain Pedro Fernandez de Valenzuela in 1555, near the Rio Somondoco, in the present region of Chivor, 90 miles (150 kilometers) north of Bogotá. Next, the near-mythic mines of Muzo were

A bove: uncut emeralds such as these plundered by the conquistadors were brought to the court of Spain aboard galleons. Often they came from Inca temples, where the gem was used in ceremonies, or from Colombia, after the seizure of King Tanja's treasury.

I n the early days of American colonization the conquistadors bartered with their Amerindian hosts. Cortés thus obtained an enormous pyramidal emerald, wider than a palm, which ornamented the palace of Texcoco. Left: the entry of Cortés into Mexico City before the fall of the Aztec capital.

discovered. These lay at the heart of a valley 60 miles (100 kilometers) east of the Colombian capital, at an altitude of 2,600 feet (800 meters) in the Andes.

There, on 9 August 1564, a Spanish settler named Juan de Penagros was galloping at

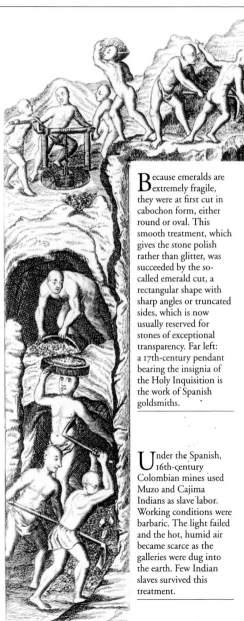

Because emeralds are extremely fragile, they were at first cut in cabochon form, either round or oval. This smooth treatment, which gives the stone polish rather than glitter, was succeeded by the so-called emerald cut, a rectangular shape with sharp angles or truncated sides, which is now usually reserved for stones of exceptional transparency. Far left: a 17th-century pendant bearing the insignia of the Holy Inquisition is the work of Spanish goldsmiths.

breakneck speed to escape a fierce battle with the Muzo Indians. Noticing that his horse was lame, he risked his life to stop and examine the horse's hoofs, and discovered, stuck in one, a superb emerald of an intense green color. The news caused a sensation in the whole region; the horseman's path was later meticulously followed backward and every

Under the Spanish, 16th-century Colombian mines used Muzo and Cajima Indians as slave labor. Working conditions were barbaric. The light failed and the hot, humid air became scarce as the galleries were dug into the earth. Few Indian slaves survived this treatment.

stone checked, until a
mountain slope rich in
emeralds was found.

This discovery should not
have been too surprising. In the
same place, long before the arrival of
the Spaniards, two formidable tribes
of Indians had clashed: the Muisca,
who possessed an emerald of exceptional
size, and who worked the local deposits,
and the Muzo, warriors who also traded
in emeralds.

The Spanish made these mines and
many others in the Americas profitable by
imposing appalling working conditions on
the Indian slaves, Muzo and Cajima, who
worked in the stifling heat and humidity of
the galleries and shafts, and died in great
numbers. Enormous quantities of emeralds
were shipped to Spain and to all Europe. To
sell all their stones the traders turned to other
markets: in the Ottoman Empire, Persia, and
India. The emerald collection of the Ottoman
sultans, largely intact in Istanbul, remains
one of the most splendid in the world.

The strange ceremony
of the Guatavia lagoon

The Americas produced a
wealth of gold and gems
after the Spanish

This magnificent
monstrance of
gold and jewels was
commissioned by
the Jesuits at Bogotà
between 1700 and
1707 and made by
Joseph Galaz. It is
called *La Lechuga*, the
lettuce, because of the
intense green of its 1,486
emeralds, which come
from the mines of Muzo
and are cut in the form
of oil-drops and
butterfly wings. The
work, considered the
masterpiece of colonial
religious goldwork, also
bears a Brazilian topaz,
62 West Indian pearls,
168 Indian amethysts,
28 South American
diamonds, 13 Ceylonese
rubies, and a Thai
sapphire.

Conquest, and with them came legends of fabulous treasures hidden in jungles and rain forests. The most famous is that of El Dorado, the lost city of gold. El Dorado has sometimes been identified with a site near the emerald fields of Chivor and Muzo, in the crater lake of Guatavia, where an initiation ceremony took place every year for the successor of the great chiefs of those peoples, amid offerings to the gods.

The future prince was stripped of clothing and his body coated with sticky clay and powdered with gold dust. He was seated in a small boat and his people laid offerings of gold and emeralds at his feet. As silence was imposed a flag was waved; then the heir, with his lieutenants at his side, threw the treasure overboard. The ceremony ended in a burst of public celebration.

Not surprisingly, the crown of Castile wasted no time in draining the lagoon in the early days of the Conquest. The operation took several years and proved fruitful, according to the Spanish chroniclers of the period; other sacred lakes were drained in their turn.

Pigeon's blood: the incomparable rubies of Burma

There is a peerless shade of red, unique in all the world, found only in the rubies that come from the famous Mogok mines of Burma (now Myanmar). The color in question is a deep, bright red, very pure, and gives the best Burmese rubies an inestimable value. No common shade of scarlet or vermilion suits these remarkable stones. We have to thank an eminent Swiss gemologist for having found the perfect description for that hue, known as "pigeon's blood."

This incomparable Burmese red appears in the twenty rubies

The 17th-century Mughal emperor Shāh Jahān wears a necklet of pearls and rubies and holds a turban ornament with an immense emerald and diamond.

(weighing up to 10 carats each) decorating a belt buckle on display at the Central Bank of Iran, in Teheran. These stones are cabochons: that is, they have smooth, unfaceted surfaces, truly like droplets of blood. This buckle is one of the most beautiful jewels in one of the world's great gem collections.

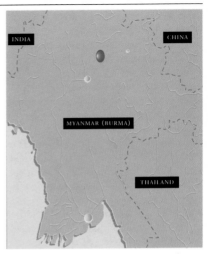

The Mogok mines were called the Valley of Rubies. They have produced the most beautiful rubies on the planet for fourteen or fifteen centuries. Tradition tells us that the king who owned the mines ordered his workers to give him the biggest stones, and left them the smallest stones as pay. This gave the workers a powerful incentive to break up the most beautiful stones in order to keep the pieces for themselves, and large Mogok rubies became very rare.

Above: the Mogok Valley lies 120 miles (200 kilometers) north of Mandalay. Below: a 17th-century Mughal white-jade flask encrusted with rubies and emeralds.

The theft of the French crown jewels

In Paris, one sunny Monday in autumn 1792, a man named Paul Miette went to admire the crown jewels of France. Though he was a famous bandit, he did not worry about being recognized or caught. The jewels had found a safe haven from the turmoil of the Revolution—at least, so it was thought—the year before in the Garde-Meuble, or royal furniture warehouse, at the Place de la Révolution in the center of Paris.

The most important of the crown treasures were kept there: collections of rare gems, arms and armor of past monarchs, tapestries, and furniture—all of incalculable worth. The guard was badly organized and often not relieved on schedule. On several occasions the general steward of the king's household had been concerned enough to warn Jean-Marie Roland de La Platière, minister of the interior.

During the night of 11 September, at around 11 o'clock,

two gangs met on the premises, one led by Miette, the other by his lieutenant, a man named Depeyron; their accomplices were posted as lookouts in the great square outside, while the two leaders climbed to the second floor of the Garde-Meuble and cut out the panes of a window with virtuoso skill. Once inside, they broke open the display cabinets containing jewels in splendid settings and filled their pockets. That night they did not touch the unmounted gemstones.

On the following nights, with great daring, the robbers returned. By candlelight they ransacked a chest of drawers that held a number of famous gems, among them the Sancy and de Guise diamonds and eighty-two Oriental rubies. These were extremely rare; at the time rubies were often confused with another red stone of lesser value, called spinel, or Balas ruby. One of these was a great ruby of some 24 carats, described in the collection's catalogue as "a large, broad Oriental ruby, lyre-shaped, pink-colored, having many icy flecks and a notch on the bottom, weighing 22¾ carats and appraised at 25,000 *livres*," a considerable sum. Some of these rubies were later found on the banks of the Seine River, at the spot where the thieves had divided the loot.

In the course of this burglary a great number of jewels disappeared forever, but by luck—or fate—Miette and his acolyte overlooked several. One of these was the Great Sapphire of Louis XIV, one of the most beautiful sapphires in the world, called the "third stone in the king's use." A trader named Perret had brought it to the Sun King, who purchased it on the spot, as he bought all rare and beautiful stones, "for the glory of France." Perret had obtained it from a German prince for the sum of 6,800 *livres*. Prior to that it had belonged to the Ruspoli family, illustrious Romans who for a time gave their name to the stone, and who, in turn, had bought it from Venetian traders. According to

The former Garde-Meuble in Paris, where the famous jewel theft took place in 1792, today houses the offices of the Ministry of the Navy.

This clip in the form of the heraldic Polish Eagle is from the collection of Louis XIV. The big central stone is a zircon; the 150 smaller red stones are table-cut rubies.

This badge of the Order of the Holy Ghost, set with 400 brilliants in silver, was given by King Louis XV (1710–74) to a member of the family of the Bourbon dukes of Parma. At center is a dove, head downward, with outspread wings and a ruby beak. A diamond of 7.5 carats forms the bird's body.

Below: the Hortensia, a pentagonal pink diamond of 21.32 carats, was cut by order of Louis XIV around 1678.

an erroneous story the stone was said to come from Bengal, which produced many of the world's beautiful colored stones at that time. This sapphire is of a magnificent color and purity. The French crown catalogue describes it as "an Oriental sapphire of good color, clear and bright, in a long form with six plane facets, more deeply colored on the ends." During the French Revolution it was pawned to finance the wars of the Directoire (1795–99) and was later donated to the National Museum of Natural History in Paris, where it is housed today.

The sapphire passport of Marco Polo

The Venetian trader and adventurer Marco Polo (1254–1324), who traveled from Italy to the heart of Asia in the 13th century, tells in *The Book*

Far left: the Petit Sancy, a 35-carat diamond in the shape of a pendant pear, ornamented the crest of the crown of Queen Marie de Médicis (1573–1642), wife of Henry IV of France, for her coronation in 1610. She had acquired it in 1604.

The coronation of King Louis XV of France took place on 25 October 1722. The king's crown (left and below) included 16 emeralds, 16 sapphires, 16 rubies, 16 topazes, 230 pearls, and 161 diamonds, among which were two exceptional gems: the Regent and the Sancy. After the ceremony the king had the real stones replaced with copies and bequeathed the crown to the treasury of the royal abbey at Saint-Denis.

of Marvels, the account of his adventures, that he carried sapphires as calling cards, as an entrée to the court of the Great Khan, the Mongol emperor. Kublai Khan (1215–94) so coveted these blue gems, emblems of immortal sovereignty, that in gratitude he paid Polo for the stones "twice what they were worth" and named him a roving ambassador.

The sapphires that the clever explorer carried to Mongolia came from the mines of Ceylon (now Sri Lanka), an island off the coast of India. These mines, which are described by Polo, have produced magnificent sapphires for well over two thousand years, found in alluvial deposits. They have always been surrounded by mystery: we do not know the precise source of the precious stones, but only that river water carries them from high in the mountains to the center of the island, to the area around Ratnapura, a city whose romantic name means "city of gems" in Sinhalese. The gravel from the riverbed is collected by miners using age-old techniques: washing and panning in large baskets of woven bamboo, as is done with gold.

In his *Natural History,* the French naturalist Georges-Louis Leclerc, Comte de Buffon (1707–88), records a passage about these mines taken from a 17th-century *History of Ceylon* written by a certain Captain Ribeyro. This text reported that in 1701 the riverbanks of Ceylon provided an abundance of colored precious stones, and that "the Moors, setting nets in the stream to catch them, find topazes, sapphires, and rubies, which they send to Persia in exchange for other goods."

Left: the 135.80-carat Great Sapphire of Louis XIV comes from Ceylon (Sri Lanka), as its inclusions indicate. Of a pure, limpid blue, it was long considered the most exquisite sapphire in the world. Its simple cut, a flat polishing of its six faces, has preserved the stone's natural beauty.

The lapidary's art

Though colored gemstones have long been prized and polished as cabochons to bring out their luster, they have

not always been cut into sparkling facets. It is hard to say with certainty when gems were first so cut. The earliest stones that show signs of polishing are the Indian diamonds that arrived in Europe in the first centuries after Christ. These are polished in small, irregular facets. Indian stone polishers used abrasion to shape and polish stones, as they still do today.

Once the techniques for faceting developed, various cuts and forms evolved, each successive style designed to bring out color and glitter in the jewel. Table-cutting—with the uppermost surface flat—was an

For centuries the West dreamed of the fabulous Orient and its untold riches. Above: workers search for precious stones in a riverbed in the mountains of Ceylon in this illustration from a 15th-century edition of Marco Polo's best-selling *Book of Marvels*, written around 1300.

Left: the superb 563.35-carat Star of India is the world's biggest star sapphire.

early popular style, developed during the Renaissance in Venice and Milan, where craftsmen worked in celebrated workshops in the service of the monarchs of all Europe.

The knowledge of fine gem cutting soon spread to Bruges, Antwerp, and Paris, and then to Lisbon and London. Table-cut, point-cut, and heart-cut diamonds are described in France in the inventories of the treasury of King Charles V (1337–80). These must have been the work of the stone polishers and diamond cutters of Paris, who were considered the greatest practitioners of the art, while by the 15th century the art of faceting itself stood at the top of the hierarchy of the crafts. The secrets of the Parisian master artisans, alas, have been lost to us. We must await the appearance of a famous diamond cutter from Bruges, Lodewyck van Berkem, in the late 15th century, to learn the mysteries of the craft at last.

Early diamond cutters sought to maintain the full size of a stone. Gradually, however, they began to choose brilliance at the

Left and right: traditionally, Indian lapidaries polished stones with grindstone and bow-saw, working by eye and keeping their facets simple to retain the full size of the stone. Tastes in gem cutting vary from place to place and era to era; most Indian stones were recut for greater sparkle when they were imported to the West.

expense of weight. The techniques of faceting and polishing became more refined; as they did, the art of the jeweler became more highly skilled as well. Gem cutting became the art of slicing and defracting light itself. The trick was to cut the stone so that the greatest possible amount of light entering it through its table, or top flat surface, should be reflected within it and back to the viewer in multiple rays.

The rose cut, which first appeared around 1500, is one of the first examples of this evolution in cutting. It is a round stone with twelve facets and resembles the bud of a newly opened rose. The Mazarin cut came next, a cushion-shaped cut (that is, squarish or oblong) with thirty-four facets. It is named for Jules Mazarin (1602–61), a French cardinal and statesman under Louis XIV who built up the fabulous treasury of the French crown by buying Tavernier's most beautiful gems.

Antwerp: the passion for diamonds

"No one may buy, sell, pawn, or give away any false stone, whether it be an imitation of diamond, of ruby, of emerald, or of sapphire, under penalty of a fine of 25 ducats, of which a third will go to the king, a third for the city and a third for whoever reports the crime." This ordinance was issued by the authorities of the city of Antwerp in 1447. It proves that commerce in precious stones was then flourishing in the Low Countries. Antwerp, after Bruges, became the leading diamond center in 15th-century Europe. This spectacular rise is explained in part by the encouragement of the ruling dukes of Burgundy, immensely wealthy and fascinated by diamonds, who did all they could to develop the trade and craft of precious stones in their Flemish possessions.

In the early 16th century the Portuguese gained control of Goa, in India, and eventually many of

It was not until 1465 that the trade of lapidary—gem worker—appeared in the archives of the city of Bruges, center of the European gem trade. Robert de Berquem, a Parisian jeweler originally from Flanders, published a treatise on precious stones and pearls in 1661 entitled *Merveilles des Indes occidentales et orientales* (*Marvels of the East and West Indies*), in which he claimed that the brilliant cut was invented in Bruges by his grandfather, Lodewyck van Berkem. Opposite above: this 1551 painting on parchment by Hans Muelich illustrates the various cuts for colored stones and diamonds. Over the centuries, the art of gem cutting has grown ever more precise, so that even fragile and brittle stones may be cut in ways that enhance their luster.

the important Indian diamond centers, including the famous mines of Golconda. They soon began to import enormous quantities of diamonds to Europe, sending them to Antwerp to be cut. Orders rolled in to the stone polishers, the guilds of jewelers and gem setters, goldsmiths, and lapidaries. These were

craftsmen who did not specialize in cutting one type of stone or another, but indiscriminately cut all precious gems, until the specialized craft of the diamond cutter arose and was organized in a separate guild on 25 October 1582. This caused a revolution among the jewelers, who wielded enormous financial power, thanks to the wealth that trade in diamonds brought.

Whole districts in Antwerp became specialized in cutting diamonds. The diamond merchants lived in Meir; the guild of diamond cutters met at the Eiermarkt, the egg market, from 1642 to 1663, and conducted business in Dominicanessenstraat. Their patron saints were Peter and Paul, whose feast days were celebrated at the church of the Carmelites. Antwerp, already an important banking and shipping city, became one of the most powerful financial centers in Europe. The French Renaissance king Francis I (1497–1547), famed connoisseur of the arts, rejected Parisian diamond cutters in favor of those of Antwerp, whose techniques were universally applauded. Rich merchants flocked to the town—some from Italy, such as the Affaitati, who financed the sailing expeditions of Vasco de Gama and Ferdinand Magellan at the beginning of the 16th century; others from Portugal, such as the

Baron Simon Rodriguez d'Evora, one of the city's greatest diamond merchants at the end of that century. Still others were Portuguese Jews, who became specialists in trading and finance.

In 1631 there were 164 diamond cutters in Antwerp. But around that time a wave of emigration began to shift the commerce in diamonds out of the area, which was ruled by Catholic Spain, and toward other northern cities with a reputation for greater religious tolerance— especially Frankfurt and Amsterdam. These towns welcomed both Protestants and Jews, who fled persecution by the Spanish Inquisition. All these trading centers today maintain their ancient, established reputation for expertise in diamonds and other gems.

Opposite: the Meir district in Antwerp was the center of the international diamond trade. Diamond merchants came from throughout Europe to live and work there.

A model of a 19th-century diamond-cutting workshop.

56

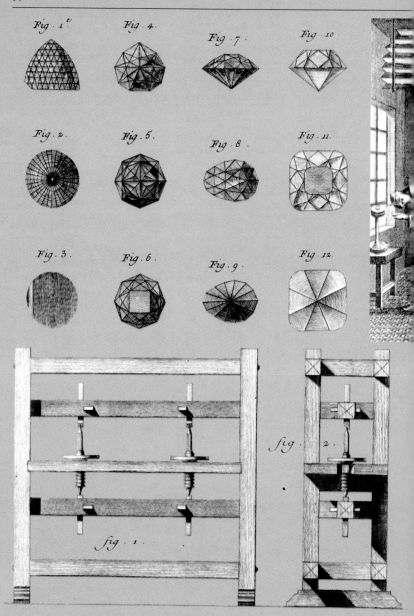

Fig. 1ᵉ. Fig. 4. Fig. 7. Fig. 10.

Fig. 2. Fig. 5. Fig. 8. Fig. 11.

Fig. 3. Fig. 6. Fig. 9. Fig. 12.

fig. 1. fig. 2.

fig. 2 *fig. 3*

The *Encyclopédie* of Denis Diderot and Jean Le Rond d'Alembert held up a mirror to the 18th century by recording the sum of human knowledge and illustrating the state of all the arts and sciences in detailed engravings. Clockwise from above left: four famous diamonds, including the Sancy (figures 7–9), are each shown in three views; stone setters at work in the shop of a goldsmith-jeweler; two men work the grindstone to shape a gem at a diamond cutter's; the grinding machine is shown in two views.

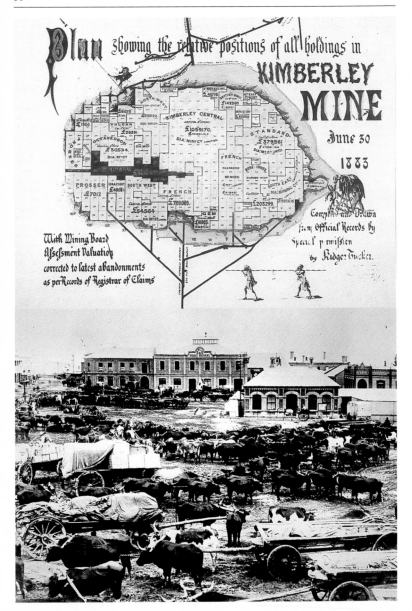

Diamond fever came to colonial South Africa in the 19th century, in the wake of the gold fever that had swept California. The fabled diamond mines of the territory proved to be immensely rich, and turned the economy of Europe upside down.

CHAPTER 3
THE AGE OF DIAMONDS

Right: the 10.73-carat Eureka Diamond was cut from a 21.25-carat rough. It was the first modern South African diamond, found by chance in 1867. After this discovery the town of Kimberley (below left) was born, an almost mythical name forever linked with the diamond. Above left: an 1883 map of claims at the Kimberley Mine Big Hole.

1867: the great South African boom

In the 1800s South Africa was colonized by both the English and the Boers, Europeans of Dutch descent who had established themselves in the Cape region in the 17th century. In the 19th century, facing increasing pressure from the former, the Boers withdrew from the coast to new regions in the free states of Transvaal and Orange. In the midst of these vast and rather arid lands, unfriendly to agriculture (except near scattered watering holes), a miracle occurred.

A Boer youth named Erasmus Jacobs lived with his relatives on the De Kalk farm by the banks of the Orange River on the outskirts of Hopetown. One day, while looking for a stick to unclog a drain, he noticed a large, shiny pebble in the mud. It was so beautiful that he brought it back to the farm to give to his sister Louisa.

He had discovered the first South African diamond, later named the Eureka. The rough gemstone weighed 21.25 carats; it came to the attention of the Colonial Secretary in Cape Town, who sent it to London to be displayed at the Exposition

After the Eureka, discovery followed discovery in South Africa. In 1869, near Hopetown, a Grigna shepherd called Swartbooi showed a neighboring farmer, Schalk Van Niekerk, a rough diamond of 83.50 carats. The farmer traded it for 500 sheep, 10 oxen, and his horse. The news spread, and thousands of fortune hunters rushed to the area. Named the Star of South Africa, the shepherd's diamond, cut in a 47.75-carat pear shape, reappeared in Geneva in 1974, where it was sold for $552,000 (£330,000).

Universelle, or World's Fair, in Paris in 1867–68. The Jacobs family refused to accept any money for the treasure, saying that a simple pebble was worth nothing.

The "Digger Rush"

After the discovery of the Eureka, other farmers found similar shiny pebbles in the area; among these was the famous Star of South Africa, which weighed 83.50 carats. The news of a major diamond deposit spread like wildfire throughout the world. In 1870 prospectors began to arrive on the banks of the Vaal and Orange Rivers by the hundreds, and then by the thousands, with only their

Right: the mining camps of the South African Diamond Rush were like little boom towns. Banks, lawyers' offices, and steam laundries opened for business, along with bars like the saloons in the American Wild West. The first to make their fortunes were the saloon keepers.

bare hands for tools. Sailors abandoned ship and rushed headlong up the 600-mile (1,000 kilometers) trail that linked the Cape with the alluvial diamond fields inland. The trip, usually made in a jolting ox-cart, took weeks. Once there, men staked claims to little parcels of land 30 feet square, pitched a tent, and anxiously elbowed one another aside to burrow into the riverbanks. Sifting pebbles one by one they made numerous finds and the excitement mounted.

At Pniel, the first diamond field in Africa, a little world of workers, merchants, crooks, and swindlers grew up around the camp, the so-called River Diggings. Here men lived in near-anarchy: a crude form of frontier justice was quickly established, with the most cunning men, many of them former outlaws, in charge. One of these was an Englishman named Stafford Parker, a former sailor and California gold prospector. He organized the diggers, who sought legal protection for their open-cut, or open-cast, mines, and proclaimed the Orange Free State.

The first diamond fields of southern Africa were found in the Orange Free State (now Free State): Dutoitspan, Bultfontein, De Beers, and Kimberley. Later came those in the Transvaal, then Pretoria and Johannesburg.

The prospectors eventually abandoned the secondary fields on the riverbanks to work their way to the primary beds a few miles distant. There diamonds were embedded in the rock of hillsides, whose watercourses washed them to the plains. Farmers sold their unfruitful land for a fortune to the diamond hunters, who now took over sites called Dutoitspan, Bultfontein, De Beers (named for the Boer farmer who owned the lands where the famous mine was dug), and Kimberley. This last field, discovered in July 1871 and named for the British Secretary of State for the Colonies, gave its name to the matrix rock in which the diamonds were found, and which is now called kimberlite.

Kimberley, the first industrial diamond mine

In 1874 there were 430 claim-holders for the Kimberley mine alone. Miners worked day and night in the Big Hole, an enormous open pit at whose bottom they sorted stones by hand—and dry, for there was no water. There was no machinery for digging or sorting, other than horse-driven windlasses used to transport ore and miners up to the surface. The camp grew rapidly and gradually turned into a town, with banks, gaming halls, churches of all denominations, grocers' stalls, houses built of wood and sheet metal (and

The town of Kimberley in South Africa gave its name to kimberlite, the principal matrix stone in which diamonds are found. Kimberlite is a very rare bluish or gray-flecked rock, composed primarily of peridotite and mica. Below: a chunk of kimberlite containing a rough diamond from the Premier mine, near Pretoria.

even of brick for the wealthiest), and canteens, which lent their name to this second wave of prospectors, known as the Canteen Rush. Miners who had made their fortunes could arrange a marriage in exchange for 25 oxen, or for gunpowder and weapons; these newly rich families constituted the first diamond society.

Between 1872 and 1874 South Africa produced more than 2 million carats of diamonds from Kimberley and nearby sites. A strange mélange of peoples worked the mines: Boer and English colonists, African natives and former slaves, and vagabonds from the four corners of the earth. As the Big Hole slowly sank into the depths of the hard, bluish matrix rock, the primitive mining techniques increasingly proved archaic, ineffective, and dangerous. Landslides and cave-ins killed many. In response, the Mine Protection Agency was created: a sort of union that collected money from all members and attempted to establish practices to prevent accidents. The miners organized themselves into professional committees and at length formed limited-liability companies to purchase and provide the heavy equipment necessary to work the mines in a more

Opposite: the various levels of the mine of Colesberg Kopje visible in this 1877 photograph correspond to the division of the land into claims, small plots which were fought over by the first diamond hunters.

Below: with archaic machinery—mere steel cables and metal buckets—blocks of kimberlite were brought to the surface from the mines. In the first days of diamond mining these machines carried men as well as ore.

modern manner. Kimberlite was now dug and transported in a raw state to the surface, where it was sifted in the open air and its gemstones extracted. Never had so many diamonds been found at one time as in those years in South Africa, and never had demand in Europe been so insatiable. Diamonds, once a great rarity, were now plentiful, and jewelry designs in the late 19th century became sparkling extravaganzas.

Every fever passes at last. Eventually the market became saturated, and in the mid-1870s the crash came: a slump in the value of diamonds that lasted twenty years. Until about 1894 the demand for stones on the world market sank precipitously and the price per carat for uncut stones fell by 60 percent. In Kimberley and the surrounding sites panic-stricken miners quickly began to sell off their claims.

The De Beers monopoly

In the midst of this brief disaster a young Englishman named Cecil Rhodes (1853–1902) founded an empire. The story of his company, De Beers, is almost synony-mous with the modern history of diamonds. The son of an Anglican vicar, Rhodes had come to South Africa for his health at age 17. He now began to acquire miners' claims, first buying up the De Beers mine and then, over the next decade, obtaining control of the French mining companies of the Cape and the Kimberley mine, which had been owned by four companies.

His diamond monopoly was born with the creation of De Beers Consolidated Mines on 13 March 1888. Ten

Left: the De Beers mine around 1880. In 1871 the De Beers brothers, the original European settlers of the mining territory, sold their mineral rights to a Dutch prospector in exchange for 25 percent of the profits. When word of the discovery of dia-monds reached Europe they were engulfed by a massive influx of fortune hunters. They sold their land, seeking a quieter home, but leaving their name to what would become the biggest diamond company in the world.

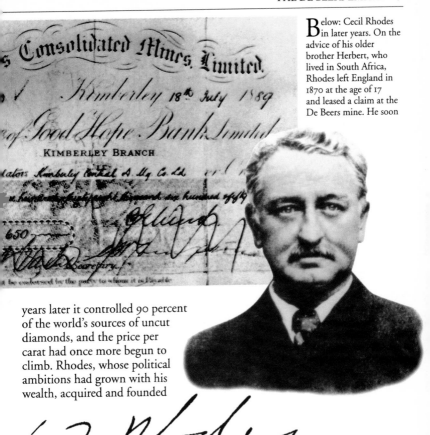

Below: Cecil Rhodes in later years. On the advice of his older brother Herbert, who lived in South Africa, Rhodes left England in 1870 at the age of 17 and leased a claim at the De Beers mine. He soon

years later it controlled 90 percent of the world's sources of uncut diamonds, and the price per carat had once more begun to climb. Rhodes, whose political ambitions had grown with his wealth, acquired and founded

advocated the consolidation of the 3,600 claims, as the best way to increase the mine's productivity. He succeeded in this and went on to buy Kimberley Central for £5,388,650 ($8,999,045) in 1889.

Above left: the famous check, preserved at the De Beers headquarters at Kimberley. Right: Rhodes's signature.

the territory of Rhodesia (now Zimbabwe and Zambia) in 1888, and became prime minister of the Cape Colony in 1890.

By 1889 De Beers had gained control of most of the South African mines. Meanwhile, the political situation in the territories had worsened: Boer resentment of English immigrants—whose numbers increased as the mines prospered—grew rapidly, and after much skirmishing erupted into a full-fledged war. As prime minister, Rhodes was a strong supporter of British

Left: the Oppenheimer family. In the early 1920s Ernest Oppenheimer founded the Consolidated Diamond Mines of South West Africa, which quickly

dominion in South Africa. He fought actively in the Anglo-Boer War, securing his mines and the De Beers company towns against the assaults of the Boers and personally directing the defense of besieged Kimberley. The war ended in 1902 with the British obtaining sovereignty of South Africa. He died that same year, immensely wealthy, and in his will established the famous Rhodes scholarship, endowed with De Beers diamond money.

The powerful De Beers empire was momentarily shaken by the discovery that year of the Premier mine, about 24 miles (40 kilometers) from Pretoria, in the north of the territory. It was here that the world's biggest diamond was discovered: the Cullinan, which weighed 3,106 carats uncut. The new mine was astonishingly rich: alone it produced as much as the entire De Beers enterprise. In addition, new alluvial fields were discovered in South West Africa, present-day Namibia, unleashing a new rush like those of the early days, notably in 1908. Periodically, word of staggeringly rich new discoveries drew successive waves of settlers, attracted by the dream of sudden fabulous wealth.

Ernest Oppenheimer reestablished De Beers's preeminence. A former broker for a big diamond

became the producer of almost a fifth of South African raw diamonds. By 1933 he virtually controlled the raw-diamond market.

Above: Lichtenburg mine, 1926: hopeful miners stake claims in a land rush.

company in London, and later a speculator in gold and copper mines, he joined the De Beers board of directors in 1926 and gained control of the company in 1929. He consolidated the business and more or less cornered the London market in uncut diamonds in 1933, in the depths of the Great Depression, by founding the Diamond Corporation Ltd (a Diamond Producer Association, or DPA), of which most African producers were members, and the Central Selling Organization (CSO), a De Beers affiliate that bought, sorted, and stocked most of the diamonds produced in the world. Together

"Within weeks, more diamonds were picked up along the banks of the Vaal River a little further north, inside the boundaries of the Orange Free State. The *Cape Argus* carried headlines that screamed *Diamonds! Diamonds! Diamonds!* and by April the tracks from Capetown were crowded with diggers of all nationalities…. One reporter said that 'they saw in their lively imaginings diamond fields glittering with diamonds like dewdrops in the waving grass or branches of trees along the Vaal River, and covering the highways and by-ways like hoar frost.'"
Graham Masterson, *Solitaire*, 1982

these associations controlled market supply.

Meanwhile, in South Africa, he was faced with continuing competition. The discovery of new deposits in the hands of new owners posed a major threat to the De Beers cartel. In 1926 alluvial beds were found at Lichtenberg, 120 miles (200 kilometers) east of Johannesburg, where thousands of prospectors produced 4.5 million carats in three years. That same year similar beds were found south of the Orange River. Angola and the Belgian Congo (now Congo) also began to produce considerable quantities of diamonds, which started to appear on the world market. Through his control of the market, Oppenheimer was able to negotiate exclusive control of the principal diamond fields.

Marketing diamonds

In 1929 came the crash of stock markets and the Depression. The Oppenheimer companies proved both efficient and patient during the worldwide economic crisis, stockpiling diamonds rather than putting them on the market at reduced prices. The name of Oppenheimer will be forever linked with the contemporary diamond trade; after his death in 1957 his son Harry took control of the business. Starting in the 1940s he had understood the value of the nascent advertising business to his merchandise. He saw that the promotion of the diamond as a luxury item must be linked to a sense of allure and mystique. In conjunction with the American advertising agency Ayer, in 1947 he created one of the world's most

Left: most of the world's uncut diamonds are bought and sold in the CSO building in London. These offices, closed to the public, are designed to enhance appreciation of the diamonds; for example, the bay windows face north, to take advantage of the purest natural light.

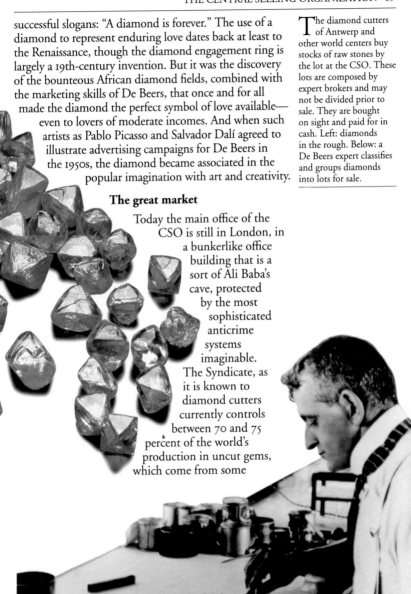

successful slogans: "A diamond is forever." The use of a diamond to represent enduring love dates back at least to the Renaissance, though the diamond engagement ring is largely a 19th-century invention. But it was the discovery of the bounteous African diamond fields, combined with the marketing skills of De Beers, that once and for all made the diamond the perfect symbol of love available— even to lovers of moderate incomes. And when such artists as Pablo Picasso and Salvador Dalí agreed to illustrate advertising campaigns for De Beers in the 1950s, the diamond became associated in the popular imagination with art and creativity.

The diamond cutters of Antwerp and other world centers buy stocks of raw stones by the lot at the CSO. These lots are composed by expert brokers and may not be divided prior to sale. They are bought on sight and paid for in cash. Left: diamonds in the rough. Below: a De Beers expert classifies and groups diamonds into lots for sale.

The great market

Today the main office of the CSO is still in London, in a bunkerlike office building that is a sort of Ali Baba's cave, protected by the most sophisticated anticrime systems imaginable. The Syndicate, as it is known to diamond cutters currently controls between 70 and 75 percent of the world's production in uncut gems, which come from some twenty countries. Diamonds are sorted into 14,000 categories, or price points, according to weight, form, brilliance, and color, and then sold to a privileged list of private clients, which De Beers guards jealously. The CSO organizes private, invitation-only sales ten times a year. There clients buy diamonds, almost sight unseen, that have been minutely described and priced beforehand by telephone. Clients arrive in person to ensure that their colossal investments correspond to their expectations and dreams. A false move is rarely made at these sales, where millions of dollars' worth of diamonds are bought, cash on delivery.

The Cullinan Diamonds

In 1903 Thomas Cullinan, an investor in diamond mines, opened a site in South Africa, within the Premier mine, near Pretoria. On 26 January 1905, at 5:00 PM, one of his workers discovered an immense diamond. The stone measured 4⅜ x 1¹⁵⁄₁₆ x 2⅜ inches (11 x 5 x 6 cm), and weighed 1.37 pounds (621.2 gm) or 3,106 carats. No bigger colorless diamond has ever been found.

Cullinan gave his name to this fabulous treasure and sold it to the government of Transvaal for £500,000 (more than $800,000), to be presented to King Edward VII of England as a birthday gift. The news of the spectacular find spread around the world. The Cullinan Diamond arrived in London by mail, wrapped in a little package, and was duly presented. To trick thieves, the attention of the media was diverted by rumors of a mysterious lead-encased package on board the royal

yacht, which was supposed to contain the stone, but which actually held a block of glass of the same size.

To design and cut jewels from the mammoth stone, the king chose the Asscher brothers, famous jewelers in Amsterdam. They had already proved their skill by cutting the 995-carat Excelsior in 1903; that stone, the second largest in the world, had also been discovered in South Africa, at the Jagersfontein mine. Because the Cullinan was so huge, and contained flaws that had to be discarded, its designers chose to cleave and cut it into many stones of different weights.

The cleaving was done in a memorable session on 10 February 1908; at the fateful moment of first splitting the stone,

Above: members of the I. J. Asscher Company, jewelers. Left: the Cullinan Diamond depicted in *The Illustrated London News* in 1908. The total weight of the 105 stones cut from the raw Cullinan (center), which weighed 3,106 carats, was 1,055.90 carats, representing a loss of 65 percent. King Edward VII added the two biggest stones, the Cullinan I and Cullinan II (below), to the crown jewels, today guarded at the Tower of London; the following year, he gave the 11.50 carat Cullinan VI to Queen Alexandra.

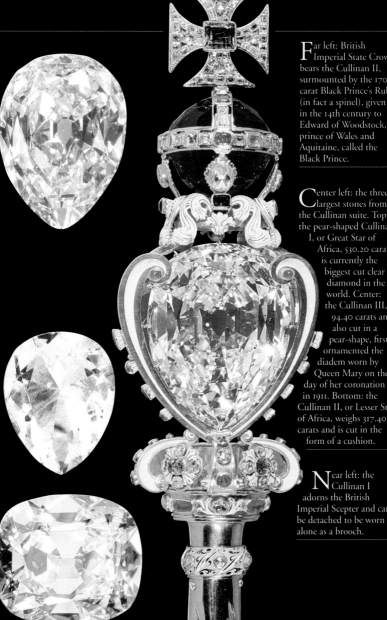

Far left: British Imperial State Crown bears the Cullinan II, surmounted by the 170-carat Black Prince's Ruby (in fact a spinel), given in the 14th century to Edward of Woodstock, prince of Wales and Aquitaine, called the Black Prince.

Center left: the three largest stones from the Cullinan suite. Top: the pear-shaped Cullinan I, or Great Star of Africa, 530.20 carats, is currently the biggest cut clear diamond in the world. Center: the Cullinan III, 94.40 carats and also cut in a pear-shape, first ornamented the diadem worn by Queen Mary on the day of her coronation in 1911. Bottom: the Cullinan II, or Lesser Star of Africa, weighs 317.40 carats and is cut in the form of a cushion.

Near left: the Cullinan I adorns the British Imperial Scepter and can be detached to be worn alone as a brooch.

Joseph Asscher fainted from stress. Nonetheless, the operation was a success and produced 9 enormous pieces (named the Great and Lesser Stars of Africa) and 96 smaller ones, all of which were faceted and polished to perfection over many months. The most important gems cut from the Cullinan are the pear-shaped Cullinan I, or Great Star of Africa, at 550.20 carats the largest polished diamond in the world, and the cushion-shaped Cullinan II, 317.40 carats. The Asscher brothers received 102 of the 105 gems as payment. In 1910 the remaining Cullinan suite was repurchased by the South African government, which then presented it to the Princess of Wales, the future Queen Mary of England. The Cullinan I was subsequently mounted in the royal scepter of Britain, and the Cullinan II is set in the British Imperial State Crown.

And there was light!

To cut the precious Cullinan stone was a matter of dizzying anxiety to Joseph Asscher—and for good reason. A diamond cutter faces a serious dilemma when planning to cut an exceptional stone. The goal is to exploit its full potential, but most large diamonds contain defects and impurities. A cutter must cleave the stone, faceting and polishing the pieces, and this entails great risks, for a

Natural diamonds are often found in octahedral crystal form. The surface of such crystals often bears slight indentations in the shape of equilateral triangles running in the same direction. These are called trigons. If crystallization has taken place slowly the crystals will be perfect, with flat, smooth planes. If changes or perturbations have occurred during growth, trigons of greater or lesser size appear. Above left and right: trigons on a raw diamond, in an enlarged photograph.

diamond may lose value if cut badly. Should the cutter try to save as much of the stone's initial weight as possible, even if its value is reduced by flaws? Or should he or she accept the loss of much material—up to 60 percent—to gain the maximum brilliance and unblemished perfection?

Cleaving diamonds is a very delicate operation, even for the most experienced diamond cutter. A diamond is split along one of the four directions determined by the internal structure of the stone, following a line that has been drawn on the surface of the stone in ink. The break is made with a special cleaving knife, which is struck with a weight. Depending on the internal structure of the stone it may also be sawed, using very precise circular sawblades made with powdered diamond—for a diamond is so hard a material that it can only be cut with a diamond. This operation demands hours of intense concentration, since the saw cuts into the stone at a rate of only a fraction of an inch an hour.

Left: because diamond is so extremely hard, it takes a lot of ingenuity —and sophisticated tools—to cut well. For centuries jewelers were limited to simple cuts: in the 14th and 15th centuries uncut diamonds were set in their original octahedral form (1). In the 16th century the table cut was developed (2), succeeded by the eight-eight cut (3), which was popular between the 16th and 17th centuries. In the 17th century the Mazarin cut (4) was invented, and shortly thereafter the Perruzi (5), fashionable until the 18th century. The old-mine cut (6) was current in the late 18th and early 19th centuries, followed by the old cut (7), which evolved into the famous brilliant cut (8), used today.

Next comes bruting, or rounding up, in which two diamonds are rubbed together to give the main stone its general shape. Finally comes faceting. This is done on a lap, a polishing tool charged, or coated, with diamond grit; each facet is cut in a precise order, so that the facets cover the whole surface of the stone equally. A 2.50-carat rough generally yields a 1-carat brilliant.

The brilliant cut

For many centuries, diamonds were simply split along the natural planes of the original octahedron of the crystal. In these older cuts a diamond had a relatively small table, or main facet, and a relatively large culet (the small facet on the underside of the stone, parallel to the table). For generations diamond cutters toiled in great secrecy in their workshops to design brighter, more sparkling cuts. Successive styles of cutting grew ever more intricate, requiring more and more faces. These culminated in the most popular modern style, the brilliant cut, whose 57 or 58 facets achieve far more dazzle than the older point, table, or rose cuts. This remarkable way of presenting a gem was probably invented in Venice in the early 17th century. It is mostly used for round stones, although it can be used on pear-shaped, marquise (oval), or heart-shaped stones as well.

It begins with the uncut stone. First the angles of the

Above left: named the Centenary, this 599-carat rough diamond from De Beers, the third-largest diamond in the world, was discovered in July 1986 in the Premier mine in South Africa. This mine is legendary for the size of its diamonds: it produced the Cullinan, as well as a quarter of all the diamonds ever discovered over 400 carats rough and some 300 stones above 100 carats rough.

An extraordinary diamond demands extraordinary treatment. The Centenary required three years of study and preparation, from 1988 to 1991, to be transformed into a 273-carat cut stone, whose unusual form is inspired by the design of a heart and an escutcheon. Left: the operation in the De Beers Diamond Research Laboratory in Johannesburg was entrusted to Gabi Tolkowsky, whose family is renowned in the industry (his great-uncle Marcel was a principal designer of the brilliant cut in 1914). The first task, which took five months using traditional methods, was to remove the undesirable pieces of the stone. Polishing took more than nine months, and the final work three months. The finished stone has 247 facets.

natural octahedron are faceted. Then the rest of the stone is faceted according to the laws of light refraction.

As diamond engagement and anniversary rings have grown in popularity, the commonest treatment for a diamond has come to be the single, solitaire-mounted, brilliant-cut stone. Such jewels come in all sizes and price ranges, and

Above: the actress Elizabeth Taylor collects diamonds. Here she wears a diadem from Van Cleef & Arpels. The actor Richard Burton gave her a pear-shaped diamond weighing 69.42 carats, set in a ring by the jeweler Harry Winston and named the Taylor-Burton.

Far left: despite the Hope's reputation for being accursed, it has never lacked for eager owners. Evalyn Walsh McLean bought the legendary 45.52-carat blue diamond from Cartier in 1910. Here she wears it on a necklet, with another famous diamond, the Star of the East (94.80 carats), as a hair ornament.

Near left: in this 1934 portrait by Boutet de Monvel, Rao III Holkar, maharajah of Indore, wears a necklet from which hang two fabulous pear-shaped diamonds from Golconda, weighing 46.95 and 46.70 carats, respectively. Winston bought them in 1946 and had them recut.

have greatly increased international demand for brilliants. As a result, the modern diamond industry works on a large scale, producing gems in immense factories around the world; those in Smolensk, Bombay, and Tel Aviv now compete with the traditional workshops of Amsterdam, Antwerp, and London.

Carat, color, clarity, cut: the 4 Cs

The quality of every stone is evaluated using four criteria, known as the four Cs: carat, color, clarity, and cut. The weight of a diamond is fixed and precise: the carat, equal to 200 milligrams, is a very ancient unit of measure, originally defined as the weight of a single carob seed (carob seeds have the property of being extremely uniform in weight). The bigger the stone, the higher the price per carat, and the greater the rarity.

The other three criteria are more subjective, and rely upon the judgment of a trained eye. A fine stone that is poorly cut loses value, as does a stone with flaws or impurities. To the lay observer most diamonds appear to have no tint, but most are tinged by a slight coloration, which is evaluated on a scale from pure white through all the colors of the rainbow. Diamonds of a pure whiteness (that is, colorlessness) are extremely rare.

The splendor of a totally colorless

Left: the Golden Maharajah Diamond weighs 65.60 carats. As its name indicates, it once belonged to a fantastically wealthy maharajah. Below: the 128.51-carat yellow Tiffany Diamond, acquired in 1879 by the renowned New York jeweler, was cut from a 287.42-carat rough found at Kimberley in 1878. Above right: the Red Raj, a unique 2.30-carat red diamond (set among a handful of white diamonds), may come from the legendary Golconda mine. It is faceted in the old-cut style, which suggests that it was mined two or three centuries ago.

diamond is unique, but should not blind us to the beauty of colored diamonds. These are ranked in order of rarity: most precious are red and green (infrequently exceeding a carat in weight), closely followed by pink, then blue and yellow (the latter are more common). Besides their extreme rarity, the passion that they inspire in some collectors inflates the price of these colored stones. While the purest white diamonds may be worth five times an off-white stone, the value of certain colored diamonds may be ten times higher. In 1984 Christie's auction house sold a pear-shaped diamond of an intense blue, weighing 42.92 carats, for $7.4 million (£4.4 million). There are only five known red diamonds in the world; one of these, weighing 0.95 carat, was sold in 1987 at Christie's for $926,315 (£554,680). The following year the 2.30-carat Raj Red from India was appraised at $42 million (£25,000).

The Condé, a pear-shaped pink diamond weighing 9.01 carats, was given to Louis II, Prince of Condé, by Louis XIII (1601–43), in thanks for his service in the Thirty Years' War. Today it is in the collection of the Musée Condé, in Chantilly, France.

Diamonds may be the most famous precious gems, but the colored stones have their own mystique. The aura of a vivid green emerald, the glow of a rare ruby, the deep gleam of a fine sapphire—each is unique. Miners from the four corners of the earth work in the most brutal conditions to drag a rainbow of gems from the mud and water, to grace the prestigious sales at Sotheby's and Christie's and to adorn kings, industrialists, and film stars with that special glitter.

CHAPTER 4
THE REIGN OF COLORED STONES

Left: the ruby and emerald owe their colors to atoms of chrome in their chemistry; sapphire's hues come from titanium and iron. Right: a fortune in rough emeralds, a common sight in Colombia.

Emeralds

Emeralds arrived in Europe in great quantities after the discovery of the New World. For the last five centuries South America has been the world's most important source of the green stone. Colombia, where mining dates back to the 10th century, produces the most beautiful gems from the mines of Chivor, Cosquez, and above all Muzo. These stones are of an intense, velvety green, neither too yellow nor too blue. At these sites quarrying is done by hand, or with limited mechanical assistance.

Bogotá, Colombia's capital, is rapidly becoming a first-class emerald-marketing center. Whole districts of the city specialize in sales, while others are dedicated to the cutting and treatment of stones. Gem workers use acids to clean fractures in the stones and remove impurities, and fill the minute cavities left after cleaning with resin, giving a stone the (purely illusory) sparkle of perfection. The work is often done on the roof of a building, so that the noxious fumes of the chemicals will escape into the atmosphere. Some of these treatments, which serious professionals abhor, shorten the life of the jewel.

As in the commercial world of diamonds, raw emeralds are sold wholesale by lots, measured by carat. Negotiations over a lot of a hundred or so carats may last for days. The market is truly international; an important sale brings buyers rushing from around the world. In Colombia mines are often privately owned. The miners work on six-month contracts and have the right to keep all the loose emeralds they may find on the ground, after they have dynamited the calcite seams to free the ore. The bigger crystals that are embedded in the blasted ore remain the property of the mine owner or franchise holder, who ships them to Bogotá for sale. Indeed, the most intrepid emerald dealers come out from the city to the mines and stand at the entrances in the early morning, to be the first to negotiate directly with the miners for the most beautiful stones.

Miners the world over follow a difficult, dangerous

Because of their terrible working conditions, the Colombian mines at Muzo are commonly known as "emerald hell." Above: downstream from the main mines and outside their barbed-wire fences, the *guaquero* scavengers work. Some 20,000 men and women sift through the muddy residue of the mined

profession. In addition to those who work the veins of the earth, the emerald industry of Colombia also relies upon *guaqueros,* a little world of outcasts—men, women, children, the old, and the miserably disabled—who gather by the thousands in the narrow Colombian river valleys, downstream from the mines. There they scavenge tirelessly in the mud for loose, stream-borne stones. They hunt day and night, huddled under insalubrious temporary shelters, ravaged by tropical humidity and insects, their faces streaked with sweat, ever hopeful of finding the *verde,* the precious green.

earth, hunting for overlooked gems. These unpoliced camps operate in a climate of desperation and suspicion, breeding much violence, and crimes are common. Within the precincts of the mines, trained workers search for emeralds in private concessions vigilantly patrolled by heavily armed guards.

Colombia's rival, Brazil

The tradition of emerald mining in South America goes back a thousand years, and Brazil too has deposits of fine green emeralds, found in the mines of Brumado, Conquista, Pilão Arcado, and Carnaiba, in the state of Bahia. Yet these are all modern fields, only in operation as mines since 1910. In spite of this late start, Brazil has recently become one of the big producers of emerald beryls, especially in the last thirty years. Most of its exports come from Bahia, or from the deposits of Minas Gerais and the states of Goiás and Espirito Santo.

Green Africa

Africa too has emerald mines. Zambia, situated in the center of the continent, between Angola and Mozambique, has produced emeralds at its Mikou mine since the 1970s. At first destined mainly for the French market, Zambian emeralds are now usually bought by Israeli merchants and cut for retail in Tel Aviv. Their bluish green color is quite unusual and is especially popular in Japan and the United States, where they command prices as high as those of Colombia.

Mines in South Africa, where deposits were discovered in 1927, yield emeralds with distinctive icy flecks. Zimbabwe, where exquisite green stones were first found in 1956–57 at the Sandawana mine, became the world's second-largest producer in the 1960s. More recently, Madagascar has joined the circle of emerald producers: in the 1990s, at the mines of Akadilana in the southeast of the island, an exceptional cluster of emerald crystals weighing almost 200 pounds (72 kilos) was found.

Rubies and sapphires

Asia is home to the biggest producers of colored stones in the world. In the traditional view of precious gems, no distinction was made between rubies and sapphires,

Israel, an important element of the gem market, buys great quantities of emeralds from Zambia for its stone-cutting workshops in Tel Aviv. More than $33.4 million (£20 million) of the green stones were imported from the central African nation in the 1980s. But emeralds from a variety of sources supply the market, including those from the Swat Valley in Pakistan and from Penshir in Afghanistan. Afghan emeralds helped to finance that country's resistance to the Soviet Union in their war in the 1980s. Above: a Zambian mine.

which were called red rubies and blue rubies. This confusion is explained by the fact that both belong to the same family of minerals; both are classed as corundums, composed principally of aluminum oxide, and both rubies and sapphires come from the same geographical areas. Corundums that possess a red color, due to the presence of chrome oxide in their structure, are called rubies; all others are called sapphires, and may be blue, violet, yellow, orange, green, or even colorless or black. Sapphires of colors other than blue are called fancy sapphires. The blue color that we commonly associate with the name sapphire comes from titanium oxide and iron.

The appreciation of colored precious stones raises a complex problem in the marketplace. How are such gems to be graded? The evaluation of the quality of hue, despite some objective criteria, is necessarily a matter of taste, to a degree. Unlike diamonds, whose color is rated according to its clarity and purity, the

These stones—blue, violet, yellow, orange, pink, and green —are all sapphires. Some colors are particularly sought after and command high prices. But the values of gems are common knowledge to producers, and even professional buyers may acquire stones at discount only by haggling or by trick. One such dealer went to Sri Lanka disguised as a butterfly hunter, to scout for a pink sapphire of about 20 carats. As he pretended to look for butterflies he came across precious stones, and one day he saw the sapphire he was looking for. By seeming to be a disinterested novice, he was able to buy it for a reasonable price.

value of colored gems is based on the quality of their color, rather than their flawlessness: a jewel of delicate or·rich tint with a small defect is preferable to an unmarked stone of mediocre color. Rubies are far rarer than diamonds, emeralds, or sapphires, and are the stones most sought by collectors. Clean stones of good color are sold for such exorbitant prices that the market for them can only be compared to that for Old Master paintings. In 1988, for example, a ring set with a 15.97-carat Burmese ruby was auctioned at Sotheby's for $228,252, while in the same year a diamond ring of 52.59 carats was auctioned at Christie's for $142,232.

Myanmar, land of rubies

The most sumptuous rubies still come from the Valley of Rubies in the Mogok region of Burma, or Myanmar. Its deposits have yielded five-sixths of the world's cumulative production of rubies for fourteen or fifteen centuries. In 1962 the nationalization of Burma's mines led to a greatly reduced output. Guarding the borders with extreme rigor against the illegal export of sapphires and rubies, the authorities have sometimes preferred to allow the mines to flood, rather than risk the uncontrolled escape of the alluvial stones. Burmese raw gems are marketed in annual sapphire and ruby sales, organized by the government of Myanmar in Rangoon and closed to all but an elite corps of professionals—no more than sixty people.

Thailand at the hub of the market

Sooner or later, most of the world's sapphires and rubies pass through Thailand, and some are mined there. Thai prospectors find stones locally, but also control the market in corundums mined in Cambodia, Kenya, and Tanzania. They have concessions in those countries and select the most interesting rough stones to be sent home for preparation and sale. Corundums arrive at

The pigeon's-blood red of this famous Burmese ruby is the most highly prized shade. The stone was given to the Smithsonian Institution in Washington, D.C., in 1938 and named the DeLong Star Ruby, after the donor. It weighs 100.30 carats, and its starry reflection is an added attraction.

Chantaburi, near Pattaya, from the world over to augment local production.

The Thais, past masters in the art of brokering gems, have organized an extremely simple market in Bangkok, held on sidewalk tables, where owners and miners come to sell their roughs. Thai merchants often work through Chinese intermediaries, who buy the roughs and resell them to artisans. These workers in turn heat the stones, which improves their color, and then sell them to stone cutters to be finished. The stones are heated in ovens to temperatures of 1800° F (1000° C) or more; a stone of poor color can thus be given an exceptional color, which will never fade.

Around the Mogok mines in Myanmar, every morning at about 5 o'clock, merchants spread their wares on tables. Though the country is heavily militarized, sales are not usually monitored.

Left: many rubies and sapphires are sold on the black market and escape official controls, despite some military supervision of the gemstone trade. Small dealers and brokers often work in the unauthorized markets at Taunggyi, the capital of Shan state, and Nyaungu, south of Mandalay. When they have big lots of stones they may risk slipping into Thailand, to sell their fine stones at the plentiful, active markets of Bangkok, whose gems tend to be of lesser quality.

Sri Lanka, fabled Serendib

Famous for centuries, the deposits of Sri Lanka have furnished half the world's blue sapphires, as well as a not inconsiderable quantity of rubies and sapphires of other colors. There, small-scale mining still follows a tradition twenty-five centuries old. When Marco Polo visited Sri Lanka, which he called Serendib, he came away not only with gems, but with marvelous stories about them.

The famous mine at Ratnapura has produced rubies since at least the 7th century BC. The procedure for harvesting gems is ancient. As one enters the area, one immediately sees a multitude of huts made of lashed saplings covered with palm fronds. These mark the little plots of land granted by the state on the condition that they be worked. A vertical gallery dug deep into the ground gives access to the *illam*, the zone of alluvial gravel deposited millions of years ago by water rushing down from the mountains. The miners, their bodies submerged waist deep in water, collect river gravel

and pour it into a basket of woven reeds. With a rapid circular movement of the arms, they swirl the basket in the water to rinse away the earth and grit; left in the bottom is a residue of gravel. On lucky days this may include a few rubies, superb blue or fancy sapphires, and semiprecious stones. The primary deposits at the higher elevations contain rubies and sapphires embedded in matrix rock; lower down, secondary deposits bear gemstones transported long ago by rivers and dropped in their beds.

The remarkable stones that haunt the dreams of collectors are of course extremely hard to find. Prime among these are the dazzling padparadshah sapphires, whose name is a Sinhalese word (from a Sanskrit origin) for lotus, since their color is the extraordinary pinkish orange of the lotus flower, and the spectacular Sri Lankan star rubies and sapphires, cut in cabochon form to reveal a shimmering star in the jewel's depths.

Kashmir

Though they rarely come on the market, the bright, velvety, cornflower-blue sapphires of Kashmir are incomparably splendid, a connoisseur's treasure. The first stones were discovered by chance by a hunter in 1881, in the remote Kudi Valley, at an altitude of more than 13,000 feet (4,000 meters), 35 miles (60 kilometers) on foot from the closest town. For a time the local inhabitants bartered sapphires for their weight in salt. The high altitude and severe climate restrict mining in Kashmir to

Opposite: surrounded by a little mountain range in the heart of Sri Lanka, the City of Gems, Ratnapura, is an active trading center. Here, a prospector sorts gems using a traditional reed basket.

This magnificent, extremely rare padparadshah sapphire weighs 100 carats and has the pink-orange coloring prized by connoisseurs, Such stones are found only in Sri Lanka.

at most one or two months a year. The mines, more or less abandoned, are today thought to be largely worked out.

Other sapphire-producing countries

Australia is currently the world's leading producer of sapphires and mines stones on a massive industrial scale. In contrast to the traditional, labor-intensive techniques used in Sri Lanka, Australian mines employ enormous machines, made necessary by the immensity of the territory. The Anakie mines in Queensland produce blue-black sapphires in astonishing record numbers. Other deposits produce more attractive stones: deep blue at Inverell and royal blue at Glenn-Inness in New South Wales. Most of these stones are exported to Bangkok, where they are cut and resold.

In Africa, the Umba Valley in Tanzania and Andranomdahbo, on the island of Madagascar, have recently begun to produce sapphires. A herdsman named Gaston found a very fine 40-carat sapphire of an exceptional blue there in 1990. This discovery led to a flood of prospectors throughout the island. Today production has dropped greatly, and the prospectors have moved to the north of the country,

The stones mined in the south of Madagascar are not of notable quality; they often have variations in color and are often of a blue-green tint. On the other hand, their large size makes them attractive for export to the large and busy Thai market, where they are heat-treated to "improve" their color. Many of these stones are then sold on the world market, without mention of their origins.

In Madagascar the mail system is not well-established, and it is common practice to send stones wrapped in a banknote for security.

where new deposits have been discovered in the region of Ambilobe.

Taste, price, and color

The nuances of the colors of precious stones vary, depending on their origins. Only professionals, however, can readily identify a stone's provenance, distinguishing a Burmese sapphire from a Thai or Australian stone, or a lesser-quality Burmese ruby from a beautiful Pakistani stone of intense red. Despite the public's love of colored

The mine at Andranomdahbo in Madagascar has seen thousands of prospectors stream through its sumptuous baobab forests. The miners there have dug the earth into myriad tunnels, pits, and galleries, like human moles.

stones, and fascination with the legends attached to their near-mythical sources, leading jewelers are rarely moved by such tales, and prefer, for example, a high-quality Sri Lankan sapphire to a guaranteed Kashmiri sapphire of lesser quality, or a good Thai or Pakistani ruby to one from exotic Myanmar.

Colored gems display infinite variations in color, brightness, and tonality, and truly exceptional specimens are extremely rare. Thus, a jeweler wishing to create a necklace or bracelet of matched stones must exercise a great deal of diligence, patience, and perseverance and must possess a perfect knowledge of the sources, as well as relying on luck. Such jewels are not assembled overnight. In 1959 the jewelry firm of Van Cleef & Arpels began searching desperately throughout the whole world for twenty-two perfectly matched emerald-cut emeralds to make a sumptuous choker set off with diamonds; the suite was completed only in 1961. In 1958 the same firm spent months collecting seven Burmese pigeon's-blood rubies for a choker, and was forced to break up a pair of earrings to complete the piece.

Carat Diamond
ダイヤモンドは永遠の輝き
De Beers

Long the prerogative of the powerful, will precious stones become available to all, as this Japanese advertisement for De Beers suggests? Left: the maharajah of Sind, decked with emeralds; below: a 152.35-carat cabochon sapphire from the collection of the duchess of Windsor.

The future for the market in precious stones

Despite periodic or long-term economic crises in many of the developing countries that produce, sell, and buy gems, the market for diamonds continues to expand. The figures for sales of jewelry and gemstones published by the Central Selling Organization have grown from

$22 billion (£13.17 thousand million) in 1985 to $48 billion (£28.74 thousand million) in 1995. In particular, sales of cut diamonds have increased steadily, thanks in part to the development of new Asian markets in China, Vietnam, Cambodia, Malaysia, Indonesia, and the Philippines.

The situation is different for colored precious stones, whose markets and production are less well organized than those of the diamond, and whose mining remains small-scale, for the most part. In addition, many of the most beautiful gemstones come from countries whose political instability and difficult terrain hinder production and make the market both inefficient and erratic.

Overleaf: an Indian gem cutter at work.

Empress Farah Dibah Pahlavi of Iran on the day of her coronation.

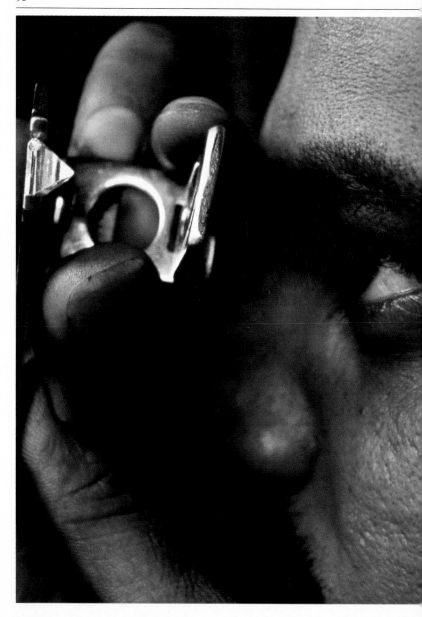

DOCUMENTS

Merchants and travelers

Since ancient times, the principal source of precious stones has been the Orient. In the 13th century the Venetian merchant Marco Polo gave the West a marvelous account of the places where diamonds, rubies, and other gems were mined. Four hundred years later the descriptions of the Frenchman Jean-Baptiste Tavernier substantiated the most fantastic legends about the riches of the Indies. In our own century, Joseph Kessel is still subject to their fascination.

A bove: mining rubies from the mountain of Sighnan, scene from a French edition of Marco Polo's *Book of Marvels*. Previous page: a manuscript illumination shows men gathering gems.

Marco Polo

Polo is still the most famous of all international travelers and adventurers. In 1271, at the age of 17, he left his home in Venice with his father and uncle and journeyed to the far ends of the known world—to Zanzibar, India, China, Japan, and Java.

You must know that rubies are found in this Island [Sri Lanka] and in no other country in the world but this. They find there also sapphires and topazes and amethysts, and many other stones of price. And the King of this Island possesses a ruby which is the finest and biggest in the world; I will tell you what it is like. It is about a palm in length, and as thick as a man's arm; to look at, it is the most resplendent object upon earth; it is quite free from flaw and as red as fire. Its value is so great that a price for it in money could hardly be named at all. You must know that the Great Kaan sent an embassy and begged the King as a favour greatly desired by him to sell him this ruby, offering to give for it the ransom of a city, or in fact what the King would. But the King replied that on no account whatever would he sell it, for it had come to him from his ancestors, and by rights he should leave it to his sons and his descendants, for he considered this jewel as a very great symbol of his reign. With that response, and without the ruby, the ambassadors returned to their master and I, Marco Polo, was one of the ambassadors: I saw the said ruby with my own eyes....

And I will tell you another thing: no man may take any large or valuable stone from his kingdom, nor any pearl which weighs a demi-*saggio* and above, unless he finds a way to smuggle it out. Thus the king does because he wants them all for himself. It is true that each year he has it proclaimed and published several times in his kingdom that all those who have beautiful pearls and handsome stones should bring them to the Court, and he will give them double of what they are worth. And it is the custom of the Court to give double what good stones cost; and the merchants, and all other folk, when they have good stones, bring them willingly to Court, because they will be well paid. That is the reason for which that King has such riches and valuable stones....

It is in this kingdom that diamonds are got; and I will tell you how. There are certain lofty mountains in those parts; and when the winter rains fall, which are very heavy, the waters come roaring down the mountains in great torrents. When the rains are over, and the waters from the mountains have ceased to flow, they search the beds of the torrents and find plenty of diamonds. In summer also there are plenty to be found in the mountains, but the heat of the sun is so great that it is scarcely possible to go thither, nor is there then a drop of water to be found. Moreover in those mountains great serpents are rife to a marvellous degree, besides other vermin, and this owing to the great heat. The serpents are also the most venomous in existence, insomuch that any one going to that region runs fearful peril; for many have been destroyed by these evil reptiles.

Now among these mountains there are certain great and deep valleys, to the bottom of which there is no access. Wherefore the men who go in search of the diamonds take with them pieces of flesh, as lean as they can get, and these they cast into the bottom of a valley. Now there are numbers of white eagles that haunt those mountains and feed upon the serpents. When the eagles see the meat thrown down they pounce upon it and carry it up to some rocky hilltop where they begin to rend it. But there are men on watch, and as soon as they see that the eagles have settled they raise a loud shouting to drive them away. And when the eagles are thus frightened away the men recover the pieces of meat, and find them full of diamonds which have stuck to the meat down in the bottom. For the abundance of diamonds down there in the depths of the valleys is astonishing, but nobody can get down; and if one could, it would be only to be incontinently devoured by the serpents which are so rife there.

There is also another way of getting diamonds. The people go to the nests of those white eagles, of which there are many, and in their droppings they find plenty of diamonds which the birds have swallowed in devouring the meat that was cast into the valleys. And, when the eagles themselves are taken, diamonds are found in their stomachs.

So now I have told you three different ways in which these stones are found. No other country but this kingdom of Mutfili produces them, but there they are found both abundantly and of large size. Those that are brought to our part of the world are only the refuse, as it were, of the finer and larger stones. For the flower of the diamonds and other large gems, as well as the largest pearls, are all carried to the Great Kaan and other Kings and Princes of those regions; in truth they possess all the great treasures of the world....

Marco Polo,
Livre des merveilles (The Book of Marvels;
also called *The Travels*), c. 1300,
translated by Henry Yule, 1903

Jean-Baptiste Tavernier

Tavernier, a gem dealer by trade and a prolific writer, has left us detailed information about how his business worked in the 17th century. The following excerpts are from his book Voyages des Indes (Travels in India), *published in 1681, part of his larger text,* Les six voyages de J.-B. Tavernier (The Six Voyages of J.-B. Tavernier). *Here he describes the Mughal emperor's famous Peacock Throne.*

VOYAGES DES INDES·
LIVRE SECOND.

DESCRIPTION HISTORIQUE
& politique de l'Empire du Grand Mogol.

CHAPITRE PREMIER.

Page from an early edition of Tavernier's book.

The principal throne, which is placed in the hall of the first court, is nearly of the form and size of our camp beds; that is to say, it is about 6 feet long and 4 wide.... Both the feet and the bars [of the base of the throne], which are more than 18 inches long, are covered with gold inlaid and enriched with numerous diamonds, rubies, and emeralds. In the middle of each bar there is a large *balass* ruby, cut *en cabuchon,* with four emeralds round it, which form a square cross. Next in succession, from one side to the other along the length of the bars there are similar crosses, arranged so that in one the ruby is in the middle of four emeralds, and in another the emerald is

in the middle and four *balass* rubies surround it. The emeralds are table-cut, and the intervals between the rubies and emeralds are covered with diamonds, the largest of which do not exceed 10 to 12 carats in weight, all being showy stones, but very flat. There are also in some parts pearls set in gold, and upon one of the longer sides of the throne there are four steps to ascend it....

I counted the large *balass* rubies on the great throne, and there are about 108, all *cabuchons,* the least of which weighs 100 carats, but there are some which weigh apparently 200 and more. As for the emeralds, there are plenty of good colour, but they have many flaws; the largest may weigh 60 carats, and the least 30 carats. I counted about one hundred and sixteen (116); thus there are more emeralds than rubies....

This is what I have been able to observe regarding this famous throne, commenced by Tamerlane and completed by Shāh Jahān.

Book II, chapter 8

Later, Tavernier describes a festive and lavish imperial ceremony:

On the first day of November 1665 I went to the palace for the purpose of taking leave of the King, but he said that he did not wish me to depart without having seen his jewels, and until I had witnessed the grandeur of his *fête....*

Immediately on my arrival at the Court the two custodians of the King's jewels, of whom I have elsewhere spoken, accompanied me into the presence of his Majesty; and after I had made him the ordinary salutation, they conducted me into a small apartment, which is at one end of the hall where the King was seated on his throne, and from whence he was able to see us. I found in this apartment Akil Khān, chief of the jewel treasury, who, when he saw us, ordered four of the King's eunuchs to go for the jewels, which were brought in two large wooden trays lacquered with gold leaf, and covered with small cloths made expressly for the purpose—one of red velvet and the other of green brocaded velvet. After these trays were uncovered, and all the pieces had been counted three times over, a list was prepared by three scribes who were present. For the Indians do everything with great circumspection and patience, and when they see any one who acts with precipitation, or becomes angry, they gaze at him without saying anything, and smile as at a madman.

The first piece which Akil Khān placed in my hands was the great diamond, which is a round rose, very high at one side. At the basal margin it has a small notch and a little flaw inside. Its water is beautiful, and it weighs three hundred and nineteen and a half (319½) *ratis,* which are equal to two hundred and eighty (280) of our carats—the *rati* being ⅞th of our carat. When Mîr Jumlā, who betrayed the King of Golconda, his master, presented this stone to Shāh Jahān, to whose side he attached himself, it was then in the rough, and weighed nine hundred (900) *ratis,*

B alas rubies (spinels) of various shapes, engraving from Tavernier's book.

which are equivalent to seven hundred and eighty-seven and a half (787½) carats; and it had several flaws.

If this stone had been in Europe it would have been treated in a different manner, for some good pieces would have been taken from it, and it would have weighed more than it does, instead of which it has been all ground down. It was the *Sieur* Hortensio Borgio, a Venetian, who cut it, for which he was badly rewarded; for when it was cut he was reproached with having spoilt the stone, which ought to have retained a greater weight; and instead of paying him for his work, the King fined him ten thousand (10,000) rupees, and would have taken more if he had possessed it. If the *Sieur* Hortensio had understood his trade well, he would have been able to take a large piece from this stone without doing injury to the King, and without having had so much trouble grinding it; but he was not a very accomplished diamond cutter.

Also an Oriental topaz of very high colour cut in eight panels, which weighs six *melscals,* but on one side it has a small white fog within.

These, then, are the jewels of the Great Mogul, which he ordered to be shown to me as a special favour which he has never manifested to any other *Frank;* and I have held them all in my hand, and examined them with sufficient attention and leisure to be enabled to assure the reader that the description which I have just given is very exact and faithful, as is that of the thrones, which I have also had sufficient time to contemplate thoroughly.

Book II, chapter 10

Next, he describes the mines and rivers where diamonds are found:

The diamond is the most precious of all stones, and it is the article of trade to which I am most devoted. In order to acquire a thorough knowledge of it I resolved to visit all the mines, and one of the two rivers where it is found; and as the fear of dangers has never restrained me in any of my journeys, the terrible picture that was drawn of these mines, as being in barbarous countries to which one could not travel except by the most dangerous routes, served neither to terrify me nor to turn me from my intention. I have accordingly been at *four* mines, of which I am about to give descriptions, and at one of the two rivers whence diamonds are obtained, and I have encountered there neither the difficulties nor the barbarities with which those imperfectly acquainted with the country had sought to terrify me. Thus I am able to claim that I have cleared the way for others, and that I am the first European who has opened the route for the *Franks* to these mines, which are the only places in the world where the diamond is found.

Stones seen by Tavernier in the course of his travels in Asia.

The first of the mines which I visited is situated in the territory of the King of Bijapur in the Province of Carnatic, and the locality is called Ramulkota, situated five days' journey from Golconda, and eight or nine from Bijapur....

All round the place where the diamonds are found the soil is sandy, and full of rocks and jungle, somewhat comparable to the neighborhood of Fontainebleau. There are in these rocks many veins, some of half a finger in width and some of a whole finger; and the miners have small irons, crooked at the ends, which they thrust into the veins in order to draw from them the sand or earth, which they place in vessels; and it is in this earth that they afterwards find the diamonds....

There are at this mine numerous diamond-cutters, and each has only a steel wheel of about the size of our plates. They place but one stone on each wheel, and pour water incessantly on the wheel until they have found the "grain" of the stone. The "grain" being found, they pour on oil and do not spare diamond dust, although it is expensive, in order to make the stone run faster, and they weight it much more heavily than we do....

I come to the government at the mines. Business is conducted with freedom and fidelity. Two per cent on all purchases is paid to the King, who receives also a royalty from the merchants for permission to mine. These merchants having prospected with the aid of the miners, who know the spots where the diamonds are to be found, take an area of about 200 paces in circumference, where they employ fifty miners, and sometimes a hundred if they wish the work to proceed rapidly. From the day that they commence mining till they finish the merchants pay a duty of 2 *pagodas per diem* for fifty men, and 4 *pagodas* when they employ a hundred men.

These poor people only earn 3 *pagodas* per annum, although they must be men who thoroughly understand their work. As their wages are so small they do not manifest any scruple, when searching in the sand, about concealing a stone for themselves when they can, and being naked, save for a small cloth which covers their private parts, they adroitly contrive to swallow it....

Tombs of the kings of the province of Golconda, as imagined in a 19th-century engraving.

The first time I was at this mine [Kollur] there were close upon 60,000 persons who worked there, including men, women, and children, who are employed in diverse ways, the men in digging, the women and children in carrying earth, for they search for the stones at this mine in an altogether different manner from that practised at Ramulkota.

After the miners have selected the place where they desire to work, they smooth down another spot close by, and of equal or rather greater extent, around which they erect an enclosing wall of two feet in height.

At the base of this little wall they make openings, at every two feet, for the escape of the water, which they close till it is time for the water to be drawn off. This place being thus prepared, all who are about to engage in the search assemble, men, women, and children, together with their employer and a party of his relatives and friends....

Each commences to work, the men to excavate the earth, and the women and children to carry it to the place which has been prepared as I have above said. They excavate to 10, 12, or 14 feet in depth, but when they reach water there is nothing more to hope for. All the earth being carried to this place, men, women, and children raise the water with pitchers from the hole which they have excavated, and throw it upon the earth which they have placed there, in order to soften it, leaving it thus for one or two days, according to the tenacity of the clay, until it assumes the condition of soup. This being done, they open the holes which they made in the wall to let off the water, then they throw on more, so that all the slime may be removed, and nothing remain but sand. It is a kind of clay which required to be washed two or three times. Then they leave all to be dried by the sun, which is quickly effected on account of the great heat.

They have a particular kind of basket made something like a winnowing fan, in which they place the earth, which they agitate as we do when winnowing grain. The fine part is blown away, and the coarse stuff which remains is subsequently replaced on the ground.

All the earth having been thus winnowed, they spread it with a rake and make it as level as possible. They then all stand together on the earth, each with a large baton of wood like a huge pestle, half a foot wide at the base, and they pound the earth, going from one end to the other, always pounding each part two or three times; they then again place it in the baskets and winnow, as they did on the first occasion, after which they spread it out again and range themselves on one side to handle the earth and search for the diamonds, in which process they adopt the same method as at Ramulkota.

Book II, chapters 15, 16

Tavernier, concerned to pass on his mercantile expertise, explains the proper way to price a diamond, according to weight and quality:

I do not mention diamonds below 3 carats, their price being sufficiently well known.

It is first necessary to know what the diamond weighs, and next to see if it is perfect, whether it is a thick stone, square-shaped, and having all its angles perfect; whether it is of a beautiful white water, and lively, without points, and without flaws. If it is a stone cut into facettes, which is ordinarily called "a rose," it is necessary to observe whether the form is truly round or oval; whether the stone is well-spread, and whether it is not one of those lumpy stones; and,

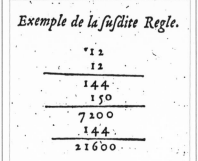

Tavernier calculates the value of a gem.

moreover, whether it is of uniform water, and is without points and flaws, as I described the thick stone.

A stone of this quality, weighing 1 carat, is worth 150 *livres* or more, and supposing it is required to know the value of a stone of 12 carats of the same degree of perfection, this is how it is to be ascertained:—

Square the 12, this amounts to 144; next multiply 144 by 150, *i.e.,* the price of 1 carat, and it amounts to 21,600 *livres*.

Book II, chapter 18

In addition to diamonds, of course, Tavernier kept records of colored gemstones:

There are only two places in the East where coloured stones are obtained, namely in the Kingdom of Pegu and in the island of Ceylon [Sri Lanka]. The first is a mountain twelve days or thereabouts from Siren in a northeast direction, and it is called Capelan. It is the mine from whence is obtained the greatest quantity of rubies, spinelles....

It is one of the poorest countries in the world; nothing comes from it but rubies, and even they are not so abundant as is generally believed, seeing that the value does not amount to 100,000 *écus* per annum.

Among all these stones you would find it difficult to meet with one of good quality, weighing 3 or 4 carats, because of the strict injunctions against allowing the removal of any which the King has not seen; and he retains all the good ones which he finds among them. This is the reason why in all my journeys I have earned a sufficiently large profit by bringing rubies from Europe into Asia.…

All the other coloured stones in this country are called by the name ruby, and are only distinguished by colour. Thus in the language of Pegu the sapphire is a blue ruby, the amethyst a violet ruby, the topaz a yellow ruby, and so with the other stones.…

The other place in the East whence rubies and other coloured stones are obtained is a river in the island of Ceylon. It comes from high mountains which are in the middle of the island, and as the rains greatly increase its size— three or four months after they have fallen, and when the water is lowered, the poor people go to search amongst the sand, where they find rubies, sapphires, and topazes. The stones from this river are generally more beautiful and cleaner than those of Pegu.

Book II, chapter 19

Tavernier made careful notes of fine diamonds and other gems that he had seen or sold in Europe and Asia. He was especially proud of his commerce for King Louis XIV of France, whom he compares favorably with the Mughal emperor:

I shall follow the order of the figures as they are arranged by their numbers, and I shall commence with the heaviest diamond of which I have any knowledge—

No. 1. This diamond belongs to the Great Mogul, who did me the honour to have it shown to me with all his other jewels.… As I was allowed to weigh it, I ascertained that it weighed 319½ *ratis,* which are equal to 279⁹⁄₁₆ of our carats. When in the rough it weighed, as I have elsewhere said, 907 *ratis,* or 793⅝ carats. This stone is of the same form

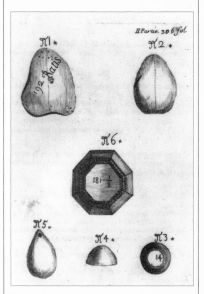

The greatest diamonds of the Mughal emperor, as recorded by Tavernier.

as if one cut an egg through the middle.

No. 2 represents the form of the Grand Duke of Tuscany's diamond, which he has had the goodness to show me upon more than one occasion. It weighs 139½ carats, but it is unfortunate that its water tends towards the colour of citron.

No. 3 is of a stone weighing 176⅛

mangelins, which amount to 242 ⁵⁄₁₆ of our carats.…When at Golconda in the year 1642, I was shown this stone, and it is the largest diamond I have seen in India in the possession of merchants. The owner allowed me to make a model of it in lead, which I sent to Surat to two of my friends, telling them of its beauty and the price, namely 500,000 rupees, which amount to 750,000 *livres* of our money. I received an order from them, that, if it was clean and of fine water, to offer 400,000 rupees, but it was impossible to purchase it at that price. Nevertheless, I believe that it could have been obtained if they would have advanced their offer to 450,000 rupees.

No. 4 represents a diamond which I bought at Ahmadābād for one of my friends. It weighed 178 *ratis,* or 157¼ of our carats.

No. 5 represents the shape of the above mentioned diamond after it had been cut on both sides. Its weight was then 94½ carats, the water being perfect. The flat side, where there were two flaws at the base, was as thin as a sheet of thick paper. When having the stone cut I had all this thin portion removed, together with a part of the point above, where a small speck of flaw still remains.

No. 6 represents another diamond which I bought in the year 1653 at the Kollur mine. It is beautiful and pure, cut at the mine. It is a thick stone, and weighs 36 *mangelins,* which are equal to 63⅜ of our carats.…

Book II, chapter 22,
Voyages des Indes (Travels in India),
1681, translated by V. Ball, 1889

How to negotiate in the Indies

In a manuscript written in the 18th century, now in the Bibliothèque Nationale de France, a textile merchant named Georges Roques describes the world of international business. His account was intended as a guide for merchants in other trades, including jewelers.

Of diamonds
There is not so much specialized knowledge required for diamonds as for pearls, although the latter display their faults while the former keep them hidden if you buy them rough. A brief apprenticeship in the lore of this jewel will be enough to make you a specialist. This is why those who trade in them should know how to shape and cleave them. For the first, remove flaws, defects and spots on the grindstone with great reluctance so as not to grind away too much of the diamond's weight, and to give it an agreeable shape, either in a rosette, table, heart, or pendant. The shape of the stone determines what cut to use. This isn't the main point: more important is to make it seem bigger and heavier than it really is. This is up to the skill of the worker. As for the knowledge about the second point, which is cleaving, that is to say to split it to get two diamonds, it isn't at all difficult since the grain of the stone corresponds to the pointed ends of the stone. It is usually in these places that they are split with a little chisel. A hammer blow, and then the lapidary polishes the results. To learn about their water and their fire, the experience of those who work with them will be the best teacher.…

Jewelry is a trade which is more refined than any other, but there have been great changes since the Mughal [emperor] has conquered the kingdom of Golconda. Since then he has closed the diamond mines for ten years by reason of the great quantity [of diamonds] that this ruler has put aside [for himself] in his treasury, to which he has added those he took from the king of Golconda, and this with the plan of getting rid of roughs, since he only loves cut stones. No one is willing to risk such trade, since they know that around his person there are diamond-traders who have the charge of this great treasure, much more wily than Europeans or Jews, from whom one can only buy stones for a great price and with great risk of being tricked. So that the Jews can only buy little lots which are found among the Banians, who inflate the prices and, since they can't use all their money on what best suits them, they only buy whatever presents itself, even merchandise from Europe or Asia.

Although my profession isn't that of diamond-trader, I have thought it best to make this little digression since a merchant should enter into commerce in whatever seems advantageous to him, but never into that which he doesn't understand.

Of the French East India Company
The Great Mughal [emperor] who, in the great extent of his empire, finds nothing in his experience which approaches his own grandeur and opulence, cannot be informed of the fine accomplishments of the French without admiration. And the proofs which have been furnished him by the little handful of those who were in his service at the conquest his Majesty has made of the kingdoms of Visiapour and of Golconda have so raised our nation above the others in his esteem that even visitors who are passing through are so highly considered by the lords of the court that there is none of them who will not eagerly procure for them from his Majesty the aid of which they are

capable in furthering the business affairs which summon them there. We have quite recently seen two good examples of this truth in two French jewelry merchants whose several purchases of diamonds and emeralds of value were detained at the customs of Surate because in the matter of stones, the Mughal demands the right to see those which exceed fifteen or sixteen *ratis* in weight which are fourteen French carats.

These Gentlemen were obliged to go to court to present them to the King to settle the matter. He only wanted to keep them with their consent and gave them in return rough diamonds which were worth double their jewels. After having overwhelmed them with the warmth of his reception and presents, he sent them back with letters ordering they not be made to pay any taxes on what they carried.

Although I report how the French are considered in the Indies, I am not at all tempted to write a panegyric on my nation. This would be repugnant, it seems to me, to the freedom of the historian to write the truth....

Only, let me add that it is only necessary to be French to obtain the grace of this prince, or to claim to be, as a Jew who speaks our language did and who, in this name, did very good business at Court and traveled, in 1690, into England and, from thence, to Holland with a very considerable share of diamonds which would have joined the treasures of the king if the title of Frenchman had not prevailed.

Georges Roques,
La Compagnie des Indes et l'art du commerce (The East India Company and the Art of Commerce),
unpublished manuscript,
18th century

The Pigeon's-blood Ruby

In a 1955 novel called The Valley of Rubies, *by Joseph Kessel, Jean, a merchant in precious stones, and Julius, his associate, pay a call on the narrator to show him an exceptional ruby.*

The man with the astrakhan hair mechanically wiped his glasses, which had thick horn rims. Then he felt in his waistcoat pocket and drew out a tiny packet wrapped in tissue paper. Jean unwrapped it neatly and with tender care. A red stone sparkled in his palm. 'Just have a look at that,' he said. He spoke in the passionate voice of a lover. 'Just look at it. It's a very rare jewel: a twenty-carat ruby, perfectly cut.'

Julius came closer. His wide face had lost a little of its Buddhist serenity.

'Pigeon's blood,' he said. 'As pure a stone as you could see…'

I didn't understand their jargon, but even through the gloom of that rainy day I could perceive clearly enough the translucent fire which smouldered with such miraculous intensity inside the crimson scrap of light.

'How much have we refused for this?' asked Jean.

'Forty-five million francs,' replied his colleague, wiping his glasses.

'Take it for a moment and hold it against your face,' said Jean to me, his loving gaze still fixed upon the jewel. 'It's warm. You can feel the life in it.'

But the moment it was between my fingers he let out a cry.

'Look out, you're going to drop it. It's fragile, you know. You'll break it.'

I was shocked. 'Do you mean to tell me that this costs over forty-five million francs and that it'll break? For heaven's sake take it away again.'

Once again the stone rested on Jean's palm. I gazed upon this incredible fortune that took up so little room.…

Julius and Jean recount the famous legend of Mogok:

'The valley of Mogok,' they said, 'lies in the north of Burma, far beyond Mandalay, between high jungle-covered hills. There have been rubies there since the beginning of time. Wrapped in their coating of rough mineral, they have lain hidden in the bowels of the rocks, along the streams, deep in the mountainsides. There—and nowhere else.

'For, as far as the memory of mankind goes back, no one has ever discovered another part of the earth's crust that will yield stones such as these—as bright as flame, as red as blood.

'Rubies are mentioned in some of the earliest writings that exist: in the Koran, in the Song of Songs, in the ancient records of India and China. Since time immemorial they have shone in the crowns and diadems of princes, kings and emperors, or lain hidden in the treasure-houses of the rajahs. Every one of them, from the most recently discovered to the oldest of all, must have come from Mogok.

'In those dim, distant times, who can have mined them, sold them, and carried them away from this wild and desolate country? No one knows: it remains a total mystery. There are magnificent legends about Mogok, but no actual records. Our knowledge of its history goes back only four hundred years.

'Yet the fact remains that, as far as it is humanly possible to discover, there can be no other geological origin for the rubies of legend and history: they must have been formed in the rocks of Mogok. And it is still this valley which supplies the world with those flame-bright, blood-red stones whose facets sparkle in the jewellers' shops of the great capitals.'

All the time Jean and Julius were telling me these things, flashes as from a red star were shining through the grey Paris light from the twenty-carat ruby—the ruby from Mogok.

At length Julius picked it up, gazed at it for a moment, and sighed.

'You don't find many like this,' he said, 'even there.'…

When they get to Mogok, Jean, Julius, and the narrator visit Daw Hla, a collector of precious stones:

Maung Khin Maung [Daw Hla's son] had brought a little packet of white paper out of the pocket of his jacket. He put it on the table and opened it, scarcely seeming to touch it with his thin, sensitive fingers. It was like a conjuring trick: there suddenly appeared, in the middle of the white paper, the first ruby I had seen in Mogok.

Jean was silent at first. His eyes, larger and brighter than usual, seemed to travel all over the little red stone, as though he were counting the facets, and absorbing its brilliance to the full. Then, without moving his eyes, he spoke in a husky voice to Julius:

'Could I have the tongs, please?'

'They're there,' said Julius, 'just under your hand.'

Jean's fingers were nothing like as delicate as those of Maung Khin Maung, but when he picked the ruby up between the slender tongs it was with the same swiftness and precision

that Maung Khin Maung would have shown. He raised it to eye level and looked at the light through it, examining it for a long time. Sometimes he held the tongs quite still, sometimes he turned them this way or that. At last he put the ruby down.

'Ah, yes, I see…' he said to Julius.

Suddenly he picked the stone up in the tongs and re-examined it.

'Come here and watch,' he said to me, 'and I'll explain about it.'

But he was really talking to himself.

'You can see how difficult it is,' he said, with feeling. 'This ruby comes within a hair's breadth of being a really fine stone—an important stone. All that area—the main part of it—is very nearly perfect, but in the background—what you might call the heart of the stone—it's all misted over. The way they were obliged to cut it prevented their being able to hide the flaw…'

He rolled the ruby slowly between his fingers; the sunshine drew fiery gleams from it.

'But just look at the rest of it,' he went on. 'It's magnificent…It's absolutely pure. It's pigeon's blood.'

'*Sang de pigeon,*' repeated Julius.

Maung Khin Maung didn't understand French, but at the last word he smiled and said in English:

'Pigeon's blood.'

And the old Chinese broker cried: 'Pigeon's blood.'

And Daw Hla herself, who knew no foreign languages, whispered in English: 'Pigeon's blood.'

It was like some mystic incantation; some magic password.…

Joseph Kessel,
*La Vallée des rubis
(The Valley of Rubies)*, 1955,
translated by Stella Rodway, 1960

Short histories of some famous gemstones

We have already met some of the world's most renowned gems: the Koh-i-Noor, the Hope, the Cullinan, the Eureka, the Tiffany, the Condé. But the gallery of great precious stones is inexhaustible. Here are the stories of some others, adapted from the accounts of two jewel historians.

The Nur-ul-Ain Diamond

The story begins in 1642: in the southwest of India Jean-Baptiste Tavernier, expert traveler and great seeker after stones, is shown an enormous pink diamond of some 300 carats, probably from the mines of Golconda. Named Darya-i-Nur, Ocean of Light (Tavernier simply called it the Great Table), the stone remained in the hands of the Mughal emperors until, in 1739, Nāder Shāh, conquered Delhi and put the city to fifty-eight days of pillage. The diamond was carried to Persia and there split in two, 176 carats for the Darya, still table-cut, 60 carats for the oval-cut Nur-ul-Ain. In the 20th century it was set by the jeweler

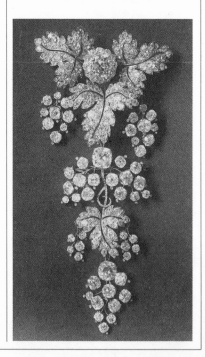

The crown jewels of France were sold by the French government of the Third Republic at an auction in Paris in 1887. Several of the most spectacular items were those of Empress Eugénie, wife of Napoléon III. Among them were a bow brooch (originally part of a larger ensemble), above, and a bodice ornament made in the form of currant leaves and berries, left, both made entirely of diamonds.

Harry Winston in a platinum tiara, amid a rainbow of multicolored diamonds, which was worn by Farah Dibah when she married the Shah of Iran in 1958.

The Black Orlov Diamond

Both its gunmetal color and its origins are enigmatic. Some say that it is the 195-carat stone, known as the Eye of Brahma, that was taken from a statue in the region of Pondicherry, in India. Others claim that it once belonged to a Russian princess named Nadia Viegyn-Orlov. Unfortunately, the said princess never existed. What's more, no black diamond is known ever to have come from India, where the color is considered unlucky. Finally, the stone's cushion cut cannot be more than a century old. Whatever the answer, the Black Orlov, which today weighs 67.50 carats, was displayed to the world as a curiosity by the New York jeweler Harry Winston before being set in a collar of diamonds and platinum and sold to a private owner.

The Vargas Diamond

Its gross weight is 726.60 carats. It is the sixth largest diamond in the world. Found in 1938 at the bottom of the San Antonio River in the state of Minas Gerais, in Brazil, it was named for the then president of the country, Getulio Vargas. As soon as Harry Winston heard of the find, he rushed to Brazil; but the stone, insured for £750,000 ($1,252,500) by Lloyds, had already been sent to Antwerp by registered mail. He ended up buying it all the same and, in 1941, the Vargas was cut: it furnished twenty-nine stones totaling 411.06 carats; the largest, an emerald-cut gem measuring 48.26 carats, was sold in Texas, then bought back by Harry Winston, who had it recut to 44.17 carats before reselling it again. In the fifties, seven other stones, from 18 to 31 carats, were mounted in a bracelet, then in a collar and assorted rings, for a maharajah.

The Dresden Green Diamond

Apple green, moss green, spring green shine from the 58 facets of the Dresden Green—40.70 carats of absolute green cut in a pendant—a freak of nature. So unique is it that we still wonder whether the stone came from India or Brazil, where mines were discovered in the 1720s. In 1742 it was bought from a Dutch merchant at Leipzig by Frederick Augustus II, elector of Saxony and king of Poland, whose father had made his capital, Dresden, a meeting place of the mind and a masterpiece of the Baroque. The jewel is known as both the Green Diamond and the Dresden Green and is a centerpiece of the treasury in the royal palace at Dresden, where it is housed in the sumptuous Green Vault. Mounted in a shoulder knot among hundreds of white diamonds, prominently displayed among the crown jewels of Saxony, the Green Diamond has no rival. It survived the February 1945 bombing that destroyed Dresden during World War II, and made a forced visit to Moscow after the war, before returning to its rebuilt city in 1958.

The Briolette of India

Legend has made Eleanor of Aquitaine the first owner of this 90.38-carat diamond. It is supposed to have been acquired in Asia Minor around 1145, at the time of the second crusade, and to have been given to her son, Richard the Lionhearted. When Richard was later captured by Henry IV of Austria, it is supposed to have paid his ransom. This scintillating diamond is mentioned in the 16th century as adorning the delicate beauty of Dianne de Poitiers, mistress of the French king Henry II. It then disappears for four centuries, but resurfaces in our era at the jewelry firm of Cartier and is sold to a maharajah. When its owner dies, Harry Winston sets it with 157 marquise-cut diamonds in a necklace and sells the lot to Dorothy Killam, wife of a Canadian financier and already owner of a 39-carat perfect blue diamond that, it is said, once graced the crown of Charlemagne. The Briolette has since been sold once again to a private individual.

The Wittelsbach Diamond

This 35.32-carat, deep blue, brilliant-cut diamond is of Indian origin; its early history is not known. It was part of the

gift of Philip IV of Spain to his daughter, Margaret Theresa, on her marriage to Leopold I of Austria in 1667.

Margaret Theresa died in 1673, and Leopold married Claudia Felicitas, who died in 1696. Leopold's third wife was Eleonora Magdalena, and he gave her all of the jewelry he had inherited from Margaret Theresa, including the great blue diamond. In 1720, Empress Eleonora Magdalena died and bequeathed the diamond to the youngest of the two archduchesses, Maria Amelia. In this way the diamond passed to the Wittelsbach family of Bavaria when, in 1722, Maria Amelia was betrothed to Charles Albert of Bavaria. It remained among the Wittelsbach jewels until 1931, when it mysteriously disappeared.

The final chapter in the history of the Wittelsbach began in a most interesting way in 1961. J. Komkommer, an Antwerp diamond dealer, was asked by another diamond merchant for advice on how to recut a large stone that he had recently acquired. On examining the 35.32-carat blue gem Komkommer felt certain that it must be of historical significance. He therefore obtained an option to buy the stone and hurried back to his office to try to identify it. After a few minutes' search he found it described and illustrated in the diamond dictionary. The lost Wittelsbach had been rediscovered. In 1964 the diamond was sold to a private collector.

The Spoonmaker

This pear-shaped diamond, weighing 85.8 carats, is one of the principal treasures of the Topkapi Museum in Istanbul. Legend has it that it was found in a rubbish heap by a Turkish fisher-

REPRÉSENTATION EXACTE
DU GRAND COLLIER EN BRILLANTS DES S.ʳˢ BOEHMER ET BASSENGE.

This engraving shows the famous diamond necklace that Marie Antoinette was rumored to have purchased with state money in 1789. The carat weights of the stones are recorded in notes. Though she did not in fact buy the piece, public anger at her spendthrift reputation fed popular resentment of the monarchy during the French Revolution.

man, who sold it to a spoon maker for three spoons. The Spoonmaker may be the Turkey II, which weighed 84 carats and was last reported in 1882 in the Turkish regalia. In the inventory at Topkapi the Spoonmaker is recorded as the Pigott Diamond. According to most authorities the Pigott was pur-chased by Ali Paşa, ruler of Albania, in 1818. Ali Paşa was an ambitious tyrant who became so powerful that in 1822 the sultan of Turkey sent an army to attack him. In the ensuing fight Ali Paşa was mortally wounded. Granted the privilege of dying in his own throne room, he ordered the destruction of his two most prized possessions, his diamond and his wife. According to the tale, the Pigott was then crushed to powder; the fate of Ali Paşa's wife is not mentioned. Another version of his story can be found interwoven in Alexandre Dumas's book *The Count of Monte Cristo,* and his court is described by the English poet Lord Byron. The possible identity of the Spoonmaker with the Pigott has been discounted by most authorities because the weight of the latter is generally given as 49 carats. However, other records variously give its weight as from 47 to 85.8 carats, so that the possibility exists that they are the same.

Extracts adapted from
Françoise Kostolany,
Le Diamant, 1992, and
Ernst A. and Jean Heiniger,
The Great Book of Jewels, 1974

A fabled jewel

Precious stones are a recurring theme in myths, legends, and fairy tales from around the world. Their symbolism is varied and rich: they may represent durability, beauty, greed, virtue, or corruption. Here, an American humorist updates the tradition with a moral tale.

The Truth about Toads

One midsummer night at the Fauna Club, some of the members fell to boasting, each of his own unique distinction or achievement.

"I am the real Macaw," squawked the Macaw proudly.

"O.K., Mac, take it easy," said the Raven, who was tending bar.

"You should have seen the one I got away from," said the Marlin. "He must have weighed a good two hundred and thirty-five pounds."

"If it weren't for me, the sun would never rise," bragged the Rooster, "and the desire of the night for the morrow would never be gratified." He wiped a tear away. "If it weren't for me, nobody would get up."

"If it weren't for me, there wouldn't *be* anybody," the Stork reminded him proudly.

"I tell them when spring is coming," the Robin chirped.

"I tell them when winter will end," the Groundhog said.

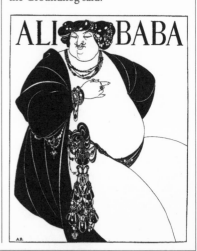

A ubrey Beardsley gave his Ali Baba jewels befitting an emperor in this 1897 illustration of the Arabian tale.

"I tell them how deep the winter will be," said the Wooly Bear.

"I swing low when a storm is coming," said the Spider. "Otherwise it wouldn't come, and the people would die of a drought."

The Mouse got into the act. "You know where it says, 'Not a creature was stirring, not even a mouse'? he hiccuped. "Well, gentlemen, that little old mouse was little old me."

"Quiet!" said the Raven, who had been lettering a sign and now hung it prominently above the bar: "Open most hearts and you will see graven upon them Vanity."

The members of the Fauna Club stared at the sign. "Probably means the Wolf, who thinks he founded Rome," said the Cat.

"Or the great Bear, who thinks he is made of stars," said the Mouse.

"Or the golden Eagle, who thinks he's made of gold," said the Rooster.

"Or the Sheep, who thinks men couldn't sleep unless they counted sheep," said the Marlin.

The Toad came up to the bar and ordered a green mint frappé with a firefly in it.

"Fireflies will make you lightheaded," warned the bartender.

"Not me," said the Toad. "Nothing can make me lightheaded. I have a precious jewel in my head." The other members of the club looked at him with mingled disbelief.

"Sure, sure," grinned the bartender, "It's a toadpaz, ain't it, Hoppy?"

A 78.86-carat diamond.

"It is an extremely beautiful emerald," said the Toad coldly, removing the firefly from his frappé and swallowing it. "Absolutely priceless emerald. *More* than priceless. Keep 'em comin'."

The bartender mixed another green mint frappé, but he put a slug in it this time instead of a firefly.

"I don't think the Toad has a precious jewel in his head," said the Macaw.

"I do," said the Cat. "Nobody could be that ugly and live unless he had an emerald in his head."

"I'll bet you a hundred fish he hasn't," said the Pelican.

"I'll bet you a hundred clams he has," said the Sandpiper.

The Toad, who was pretty well frappéd by this time, fell asleep, and the members of the club debated how to find out whether his head held an emerald, or some other precious stone. They summoned the Woodpecker from the back room and explained what was up. "If he hasn't got a hole in his head, I'll make one," said the Woodpecker.

There wasn't anything there, gleaming or lovely or precious. The bartender turned out the lights, the Rooster crowed, the sun came up, and the members of the Fauna Club went silently home to bed.

MORAL: *Open most heads and you will find nothing shining, not even a mind.*
James Thurber,
Further Fables for Our Time, 1956

Antique and modern cuts

Gems are cut in many styles. If the brilliant cut is the most common and most sparkling, other shapes and treatments are also popular: the marquise, the baguette, the lozenge, the teardrop.

1 *Round brilliant* cut (crown). The standard round brilliant consists of 58 facets: 1 table, 8 bezel cuts, 8 star facets, and 16 upper girdle facets on the crown, plus 8 pavilion facets, 16 lower girdle facets, and usually a culet on the pavilion or base.
2 *Round brilliant* cut (pavilion).
3 *Elongated cushion,* or *Indian* cut.
4 *Scissors* cut. The facets resemble scissors in an open position.
5 *Emerald* cut. A form of step cutting, rectangular in shape, but with corner facets. The number of rows of step cuts may vary, but there are usually three steps on the crown and three on the pavilion, or base.
6 *Rectangular step* cut. The step cut and the brilliant cut are the two basic classifications of cutting. In step cuts, all facets are four-sided and in steps or rows, both above and below the girdle. Although the number of rows may vary, the usual number is three on the crown and three on the pavilion. Outlines vary, but all steps are parallel to the girdle.
7 *Square emerald* cut. A form of step cutting with a square girdle outline modified by corner facets.

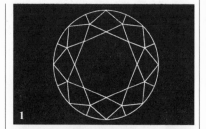

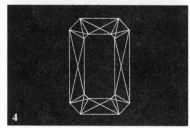

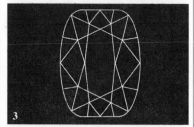

8 *Marquise* cut. The girdle outline is ship-shaped. The faceting is a modification of the brilliant cut.
9 *Lozenge* cut. A four-sided shape, usually step-cut.

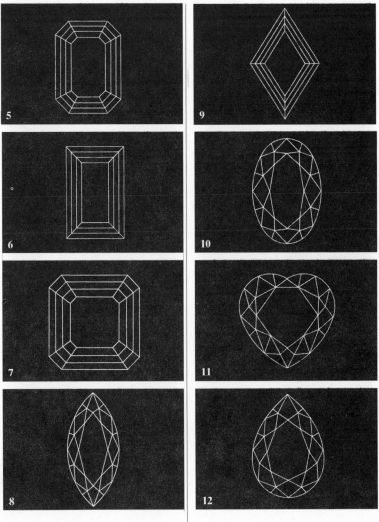

10 *Oval* shape. Based on the brilliant cut.
11 *Heart* shape. A modified brilliant cut in the shape of a heart.
12 *Pear* shape. The pear, or so-called drop-of-water shape has 58 facets and is cut in a manner similar to the brilliant.

Adapted from Ernst A. and Jean Heiniger, *The Great Book of Jewels,* 1974

The Structure of Precious Stones

	Transparency	Color	Hardness	Cleavage	Fracture
Diamond	transparent	colorless, yellow, pink, brown, black, green, blue	10	perfect	conchoidal, with splinters
Sapphire	transparent or opaque	blue, yellow, colorless, green, pink, violet, orange, black	9	none	unequal, conchoidal, splinters
Ruby	transparent or opaque	red	9	none	unequal, conchoidal, splinters
Emerald	transparent or opaque	blue-green, yellow-green, deep green	7.5	none	unequal, conchoidal, friable
Definition	*Measurable degree of clarity, lack of inclusions*	*Hues vary for each stone; colors are due to chemicals present at the time the stone crystallized*	*Measured according to the Mohs scale, from 1 to 10 (softest to hardest)*	*Degree to which a stone may be split in predetermined smooth planes along faults or weaknesses in the atomic structure of the crystal*	*Type of surface seen when a stone is shattered*

Crystalline Structure	Chemical Composition	Density	Refractive Index	Birefringence	Dispersion	Pleochroism
Cubic	Carbon	3.52	2.42	None	0.044	None
Rhombohedron	Aluminum oxide	4	1.759 to 1.778	0.008	0.018	light blue/ dark blue, colorless/orange
Rhombohedron	Aluminum oxide	4	1.759 to 1.778	0.008	0.018	violet-red/ orange-red
Hexagonal	Aluminum silicate and beryllium	2.67	1.565 to 1.600	0.006	0.014	green/ blue-green
There are seven crystalline structures for gems found in nature. Every precious stone of each kind will always crystallize in the same way	*Precious stones are crystals with distinctive atomic structures*	*The relationship between the weight of a precious stone and that of its volume of water*	*The relationship between the speed of light in the air and the speed of light in a crystal*	*Sometimes stones have two indices of refraction; such stones are called birefringent*	*The means whereby a beam of light traveling through a crystal is refracted and broken up into the spectrum at the same time*	*A property of some crystals, whose color changes, depending on the axis of observation. In a stone displaying two colors, the effect is called dichroism; three is called trichroism*

Glossary

Bezel The upper portion of a ring fitting that holds the stone; also another word for the crown of a stone

Brilliant A modern style of gem cutting with many facets (usually 58), usually on a stone with a round crown

Cabochon An unfaceted polished gem with a curved top

Carat The standard unit of weight for gem measurement; 1 carat = 200 milligrams

Cleavage The splitting of a gem along a plane, following a fault in its crystalline structure

Crown The top half of a cut gem, the part above the girdle

Culet The bottom central flat facet of the underside of a gem; parallel to the table facet

Cushion cut A variant of the brilliant cut, in an oblong or square shape with rounded corners

Cut The style or method of faceting or treating a gem to enhance its sparkle or luster

Facet The flat surface cut into a gem, or naturally occurring in the gem crystal

Gem A mineral cut, polished, or otherwise worked to intensify its beauty

Girdle The outer circumference of a cut stone, dividing the crown from the pavilion

Inclusion A small foreign body enclosed within the mass of a mineral; may be liquid, solid, or gaseous

Pavé A way of setting numerous small gems very close together, to completely cover a surface

Pavilion The bottom half of a cut gem, the part below the girdle; also called the base

Rose cut A traditional style of gem cutting with numerous facets on a stone with a round crown and a conical pavilion

Step cut A style of gem cutting, favored for emeralds, with few facets in parallel rows, or steps

Table The top central flat facet of a cut gem

Transparency The relative degree to which one can see clearly through a stone, without obstacles created by inclusions; pertains to gem-quality stones

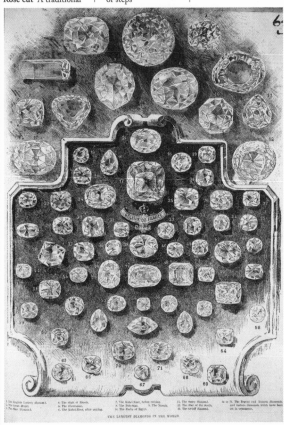

A 19th-century engraving catalogues the world's largest diamonds.

Further Reading

Arem, J. E., *Color Encyclopedia of Gemstones,* 2d ed., 1987

Bariand, P., Poirot, J.-P., Fritsch, E., *The Larousse Encyclopedia of Precious Stones,* 1992

Bruton, E., *Diamonds,* 2d ed., 1978

Cipriani, C., *The Macdonald Encyclopedia of Precious Stones,* 1986

Evans, J., *A History of Jewellry: 1100–1870,* 1970

Federman, D., *A Consumer's Guide to Colored Gemstones,* 1990

Hahn, E., *Diamond: The Spectacular Story of Earth's Rarest Treasure and Man's Greatest Greed,* 1956

Harlow, G. E., ed., *The Nature of Diamonds,* 1998

Harris, H., *Fancy-Color Diamonds,* 1994

Kanfer, S., *The Last Empire: De Beers, Diamonds, and the World,* 1993

Kunz, G. F., *The Curious Lore of Precious Stones,* rev. ed., 1989

LeGrand, J., *Diamonds: Myth, Magic, and Reality,* 1980

Post, J. E., *The National Gem Collection, National Museum of Natural History,* Smithsonian Institution, 1997

Proddow, P., Healy, D., and Fasel, M., *Hollywood Jewels: Movies, Jewelry, Stars,* 1992

Sevdermish, M., and Mashiah, A., *The Dealer's Book of Gems and Diamonds,* 1996

Sinkankas, J., *Emeralds and Other Beryls,* 1981

Sofianides, A. S., and Harlow, G. E., *Gems and Crystals,* 1990

Tait, H., *Jewelry: 7,000 Years,* 1986

Tavernier, J.-B., *Travels in India,* trans. by V. Ball, 1889

Webster, R., *Gems: Their Sources, Descriptions, and Identification,* 5th ed., rev. by P. G. Read, 1994

Wilks, J. W., and Wilks, E., *The Properties and Applications of Diamond,* 1995

Woodward, C., and Harding, R., *Gemstones, British Museum of Natural History,* 1988

List of Illustrations

A diamond-polishing workshop, recorded in a 1694 print from a book on the various skilled professions.

Queen Victoria in her official portrait for her Diamond Jubilee in 1897. She wears, among other jewels, a necklace containing 28 large diamonds as well as a pendant diamond that, together with the pendant diamond earrings, came from the treasury of Lahore, in India. They had belonged to the Timur ruby necklace, which was given to the queen by the East India Company in 1851.

Index

Acknowledgments

The author wishes in particular to thank Dominique Ardiller, Vincent Besnier, Anne de Tugny, Claire da Cunha, and Henri-Jean Schubnel. The publisher thanks Marianne Ganeau and Lesley Coldham of Diamond Information of De Beers, London.

Photograph Credits

Text Credits and Copyrights

Patrick Voillot holds a doctorate in pharmacy and
a degree from the French Institut National de Gemmologie.
Passionately interested in mineralogy since the age of eight,
he now travels the world to observe and research
gemstone mining. He is the author and editor of books
on minerals and collectable stones, and on his travels in
the places where they are found, and has also made
a documentary film on prospecting for aquamarines.

*To my friends Jean-Jacques Marcovici and Henri-Jean Schubnel,
who have helped me to realize my childhood passion.*

Translated from the French by Jack Hawkes

For Harry N. Abrams, Inc.
Editor: Eve Sinaiko
Typographic designers: Elissa Ichiyasu, Tina Thompson, Dana Sloan
Cover designer: Dana Sloan
Text permissions: Barbara Lyons

Library of Congress Cataloging-in-Publication Data

Voillot, Patrick.
 [Diamants et pierres précieuses. English]
 Diamonds and precious stones / Patrick Voillot.
    p.   cm. — (Discoveries)
 Includes bibliographical references and index.
 ISBN 0-8109-2836-1 (pbk.)
    1. Diamonds—History.   2. Gems—History.   I. Title.   II. Series :
Discoveries (New York, N.Y.)
NK7660.V6513   1998   98-7305
553.8—dc21

© Gallimard 1997

English translation © 1998 Harry N. Abrams, Inc., New York

Published in 1998 by Harry N. Abrams, Incorporated, New York

Printed and bound in Italy by Editoriale Libraria, Trieste